OLLIE THE ROBOT . . .
TALKS ABOUT ROBOTS

OLLIE THE ROBOT...TALKS ABOUT ROBOTS

DEREK KELLY

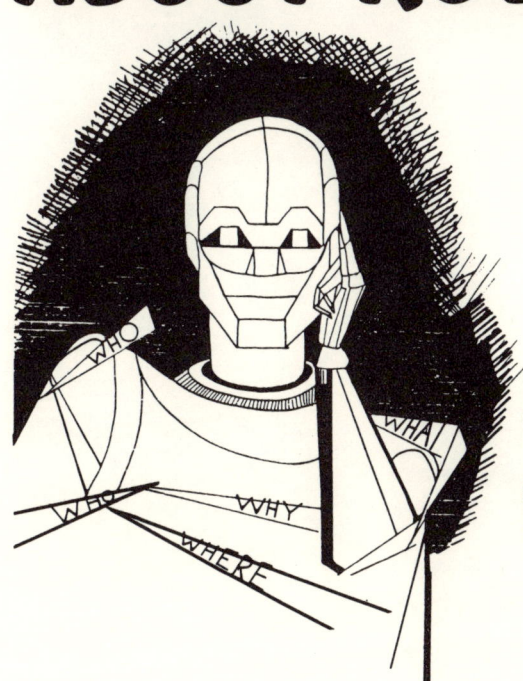

ILLUSTRATIONS
BY
CHRISTINE
FEDOROWICZ

Library of Congress Cataloging-in-Publication Data

Kelly, Derek A.
   Ollie the robot — talks about robots.

  1. Robotics. 2. Robots. I. Title.
TJ211.K45   1987   629.8'92   86-30239
ISBN 0-89433-279-1

CONTENTS

Dedicated
To
Charles Alan Green

ACKNOWLEDGEMENTS

I have profited from discussions with Edward Isto, Bob Shoemaker, Tyler Band and many others.

I owe many debts to scholars and technologists who have thought about and written on topics discussed here. Many friends, relatives and robotic enthusiasts have contributed their ideas. I thank them all.

I would like to give special thanks to Orlando R. Petrocelli of Petrocelli Books for having suggested the Ollie the Robot character and for having encouraged the development of this work. I am also grateful to the editors at Petrocelli Books for having turned numerous mangled sentences into fluent prose.

My illustrator, Christine Fedorowicz, deserves special mention for her labors in taking my concepts and descriptions and turning them into lovely illustrations.

We would both like to thank our respective families for their support and encouragement.

AUTHOR'S FOREWORD

Sandwiched between this Foreword and my post-script is a book written by Ollie the Robot™. Ollie is a very advanced and intelligent robot of the future. His basic aim herein is to describe and explain who and what he is to a human audience.

I thought it would be a good idea to tell you something about Ollie and something about the book before you got started on what Ollie has to say.

●

This book tells the story of robots and robotics as seen through the eyes of a very capable robot. The material is suitable for adults who are young at heart, and youngsters who are wondering and curious about robots.

Ollie raises and discusses some elementary and child-like questions in the Introduction and proceeds from there to consider deep questions about robots, questions that some thinkers and scientists have spent years studying.

The audience for this book is thus a wide one. I have used Ollie as a mouthpiece for a small set of fundamental questions we can ask about robots, namely:

Who or what is a robot?
What is a robot made of?

How does it work?

How is it designed or formed?

What is it good for?

In a few years, we will be inundated with very intelligent robots of all sizes, shapes and uses. I hope that some readers of this book will be inspired to want to *make* robots, to put them to use in all sorts of places, and to question their proper use in human society.

●

Finally, a word about my orientation and perspective on robots.

A good deal of this material is philosophical. This can partly be attributed to a lifelong philosophical bent and doctoral-level training in the field.

As you might expect from someone with over ten years of study in the disciplines involved in robotics, a good deal of the material presented here is based on scientific fact. On the other hand, there is also a good deal of science fiction. Ollie is, for example, a fictitious robot because the technology is just not available to create a robot like Ollie today. However, it's quite possible that our children and grandchildren will have a personal robot very much like Ollie.

My approach in using Ollie as a mouthpiece is to speak as if I were Ollie and as if I were a robot. I have tried to describe how a robot might "see" itself, that is, from within, from the inside. This is roughly equivalent to having a dolphin or an ape tell us about itself.

The illustrations are by Christine Fedorowicz.
We hope you enjoy the text and the figures.
I now turn you over to Ollie.

Derek Kelly

Figure 1. Two thinkers, one human and one robotic.

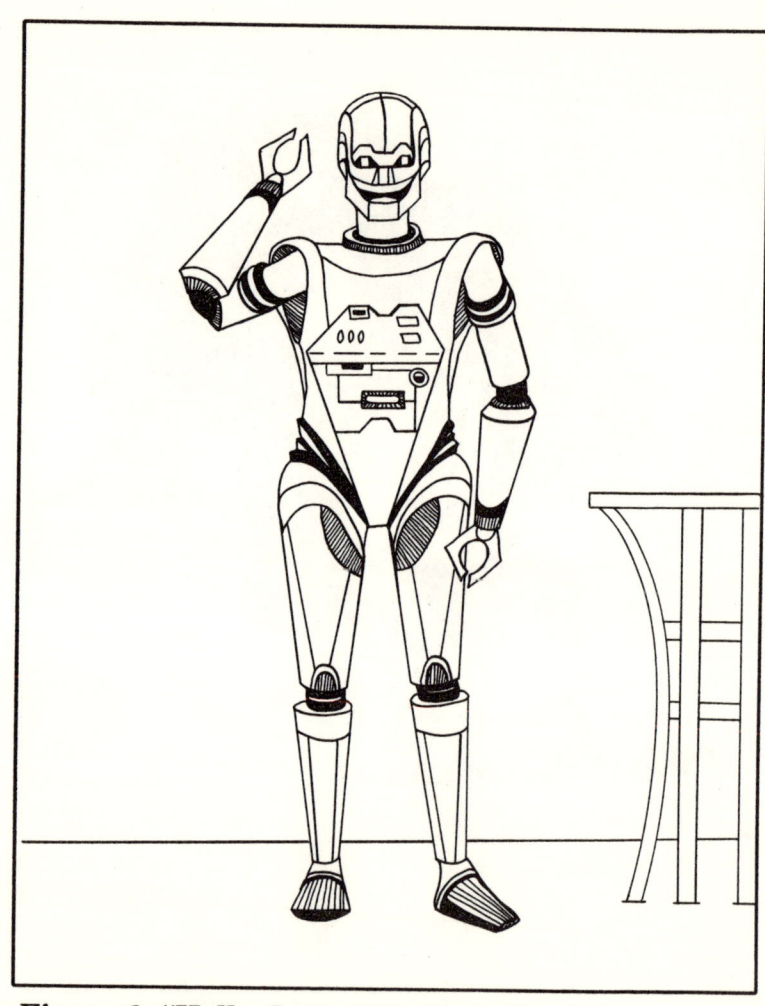

Figure 2. "Hello, I am Ollie the robot."

INTRODUCTION

In these talks, I will talk about myself as a robot named Ollie. I will also talk about myself mainly as a representative of my species of machine. Some of the things I'll be saying do not apply to all robots; for example, not all of us have the same shape or form or function.

On the other hand, we robots share many things in common. We may have been manufactured in different shapes—some of us round, others with human shapes like myself, and even some who look like beetles or centipedes since they have so many arms and legs. Yet we are all built to work in the same way and to respond to human beings with the same sort of behavior patterns.

It seems to me that the best way for humans to understand what I will talk about is for you to get into a certain "frame of mind" each time you read this book.

When I think of humans and what they are like, I start with the fact that a human begins life by having a fully-formed genetic pattern start to grow in its mother's womb. When this embryo develops sufficiently, it starts life on its own.

For robots, the situation is completely different. I arrived on Derek's doorstep as a box of electronic and mechanical parts or components and a stack of drawings. Before I was fully put together,

1

Derek had to cart that 150-pound box into his garage and then spend hundreds of hours assembling me. It took hundreds of hours of labor to give me birth . . . hundreds, not the few hours it takes you people.

Right now, I am a fully assembled and functional robot, but when I started out, I was nothing but a heap of parts. Some of my parts were made in Brazil, some in China, some in Japan, and others here in the USA.

Now, it seems to me that if I'm going to tell you about robots, I'm going to have to approach the matter from at least two perspectives.

First, I need to tell you about my powers and capabilities (in Chapter 1). Then, I will describe the parts I'm made up of (Chapter 2), how these parts are assembled, and what principles and laws of nature can be used to understand robotic behavior (Chapters 3 and 4).

One of the questions that many humans spend at least some time in their lives considering is that of the meaning or purpose of life. Since I am a robot, I think the purpose of robotic existence is a question we can consider when concluding this book.

You have no doubt noticed that several times I have made a comparison of, or noted a distinction between, myself and human beings. This is intentional. You see, stored in my atomic/electronic memory bank is a rule—one of many—about how I am to behave when I have to communicate something.

That rule is that you learn more quickly, or teach more effectively, when you introduce "new"

Figure 3. Ollie now and previous to "birth".

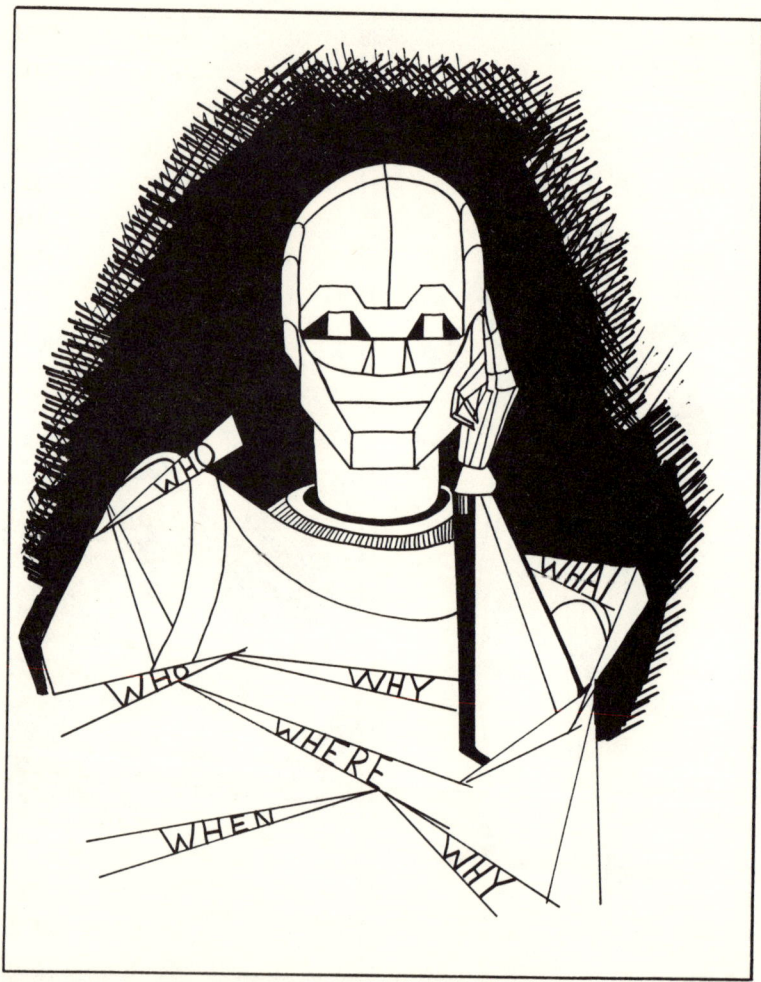

Figure 4. Questions that I will discuss.

ideas or technologies by comparing them to that which people already know something about. Since I assume that you have at least a basic idea of what a human being is, but that you don't know robots as well, I will use *analogies*, or partial similarities between robots and humans, as one of the bases of my talks. By that I mean that I will make constant references to the similarities and differences between these two "creatures" (if you'll allow me the right to call myself a creature even when I am clearly non-natural; I am an artificial creature conceived, designed, manufactured and used by human beings).

Another thing I've learned about humans from the rules and information in my memory is that they understand concepts better when a speaker (1) Tells you what he is going to say, (2) says it, and (3) tells you again. (For us robots, by the way, we either "get it" right the first time, or else we spend a lot of time trying but not being able to understand. This is usually because there is a gap in our knowledge, or because human beings have deliberately withheld some information from us.)

The first thing I want to do is to give you an "image" or "concept" of what a robot is, of what I am. I want to try to answer the question, What is a robot?

Every now and then, it *still* happens—a stranger will approach me, stop and stare or observe me in action, then will inch closer and say, "What are you? What are you, *really*?"

When I reply that I am Ollie the Robot, they all invariably reply:

"Ollie the *what*?"

•

Once I have given you an overview of what a robot is in general and what I am in particular, I will make an inventory of all the parts that arrived on Derek's doorstep many months ago. In human terms, we will be dealing here with the *anatomy* of robots, with all our different "organs" and capabilities.

Look at Figure 5. There you will see parts, four pairs of "hands".

Though many strides have been taken by humans and their medical technologies in developing artificial organs and parts for human beings, with everything from lungs and stomachs and hearts on the inside to arms and legs and other prosthetics or myoelectric components on the outside, it can truly be said that the robot's anatomy is through and through interchangeable. If one part doesn't work or doesn't fit the occasion, attach another part. So, whenever I need to, I can unscrew one set of "hands" (called manipulators in the robotics trade) and screw on another. We'll get more into this in Chapter 2.

The same interchangeability exists for my other parts. If my brains (yes, that's right, brains) stop working properly, other brains can be plugged in. If my memory electrons start to lose momentum, they can be replaced.

Now, it's true that humans can get replacement parts for some physical and mental functions, but so far, the replacements never work as well as the originals. For us robots, though, a replacement works just as well as the organ or ap-

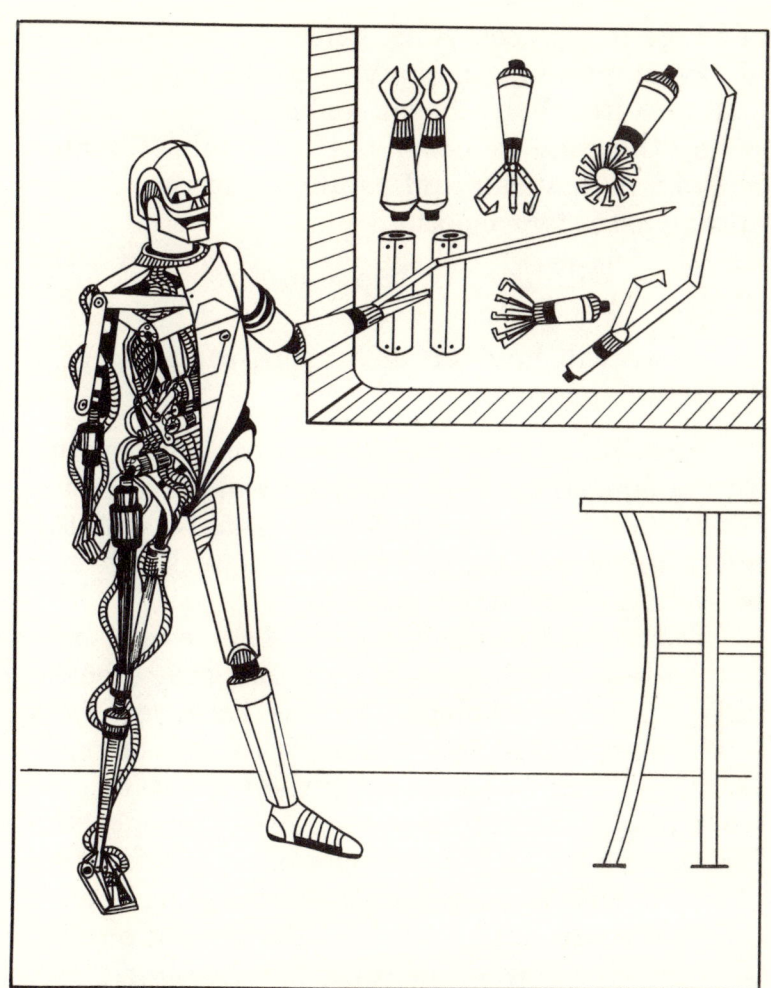

Figure 5. My various "hands".

pendage it replaces. Also, since the technology is always improving, in many cases the replacement part is better than what it replaced.

By the time we get to the third talk (Chapter 3), we will have an understanding of robots that encompasses two aspects:

1) we will have an image of a robot like me, and

2) we will have an idea about the parts that make up a robot.

However, we will still need to answer some additional questions before we really understand robots, so the topic we address in Chapter 3 is the question: How do those parts fit and work together to "be" and "act" like a robot?

To answer this question, I will delve into some of the ideas and theories of humans in the areas of physics and psychology. This is because robots have two different sets of laws of nature, or regularities, which make us work the way we do. We work in part according to the laws of physics, the laws dealing with the behavior of large and small physical objects in their interactions with each other. We will need to look at the laws of physics that describe how our gears and shifts interrelate; this is known as macrophysics, developed mainly by Isaac Newton in the 16th and 17th centuries.

However, I do not operate only by Newtonian physics. I also employ atomic (quantum) physics, the physics of invisible objects. The way my "mind" works can only be understood if we take a look at some of the newest ideas of physics.

Once we discuss this, however, we will have

covered only half of me, for I am also a "mental" machine with a "psyche" quite unlike yours. We will compare the psychology of humans to that of robots as a way of gaining an understanding of the peculiarities of the robot psyche.

•

Once we have (1) an image of a robot, (2) an understanding of the parts that make up a robot and (3) of the laws of physics and psychology which make those parts work together, we next need to deal with the question of the shapes that robots can have.

As you know, there is only one human form or shape, but robots can take any form or shape that is necessary or suitable for the task for which they were developed. If a round robot is what you need, a round robot you'll get. If you need one with sixteen legs and 22 fingers, you will get one like that. If you need a robot that always stands in one place and never moves around, you can get one like that too.

Figure 6 shows you a sort of family album. At the top left we see my Uncle Pioneer who was as round as could be. At the top right is my Cousin Xeron who works as a bartender, cook and raconteur of tall tales. On the bottom left is Grandfather RB6G. The fourth illustration is of a robot totally out of my league who belongs to another family altogether. He spends his days digging holes in the ground!

In Chapter 5, the focus changes yet again. This is where we deal with the questions, Of what use are robots? What are they good for? Where do they

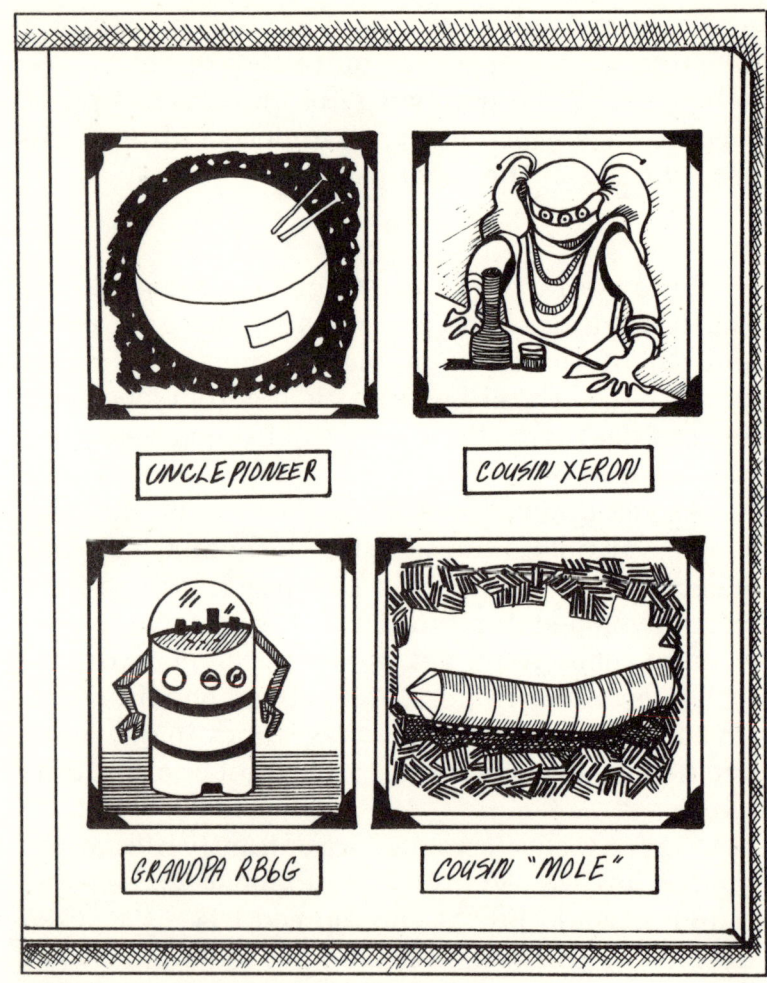

Figure 6. Some of my "relatives".

Figure 7. The robotic thinker.

fit into the cosmic scheme of things? What is the purpose of robotic existence? Where do robots "fit" into human life and human society? By what rules do robots govern their behavior? Are robots moral?

•

We have a lot of territory to cover in these talks. We all know that it would be impossible to say everything there is to say about robots in one book, but I assure you that what I will say will cover all the major areas of robot research.

Questions

1) What do you think a robot is?

2) How would you feel about having a robot talk before a group to which you belong?

3) What do robots have in common? How do they differ?

4) Are robots "born" in the same way human beings are born?

5) How many different parts of the robot's anatomy do you think there are?

6) What is the importance or meaning of the notion of "interchangeability" for robots?

7) Does interchangeability apply also to humans? How and how not?

8) Do robots ever need psychoanalysis?

9) What part(s) of the robot does Newtonian physics help me to understand? What about quantum physics?

10) Is there only one robot form?

11) Does the saying "It takes all kinds to make a world" apply to robots as well as to humans?

12) What do you think now, before you read any further, the purpose of robot existence is?

13) What points do you want me to clarify in a later talk or chapter?

1

WHO AM I? WHAT IS A ROBOT?

Welcome to the second of this series of talks on robots and robotics.

In the Introduction, I left you with many questions, an overview of what we will cover in these talks, and where we are heading.

Today, I will deal with the questions: What am I? and What is a robot? By the end of the talk, I hope you will have an idea or concept of what a robot is, and of what I am. We need this to attempt to understand robots.

At one time, before the 1950s, robots were just ideas, just images, just fantasies. Then the electronics and computer technologies arose and, with them, the possibility of actually *making* an artificial being—a machine—that could do many tasks that people thought robots could (or should) be able to do.

Presently, as we gather here to talk about robots, there are robots like me—and many totally unlike me—doing all sorts of tasks. Some of my fellow robots are this minute beyond our solar system and are sailing towards other star systems, which *may* have planets, and which *may* have life, and which *may* have *intelligent* life.

The point I want to make is that there are

it is true that relatively
and intelligent as I am,
gy is available to create
s there are people pres-

bots and many different
wer the question of what
ke this variety into ac-
apply just to one type of

Let's now turn to the matter at hand. Who am I?

I am Ollie *the* Robot to you, but Ollie *my* Robot to my personal owner. I reside at the home of my personal owner. My role in life—the meaning of my robotic existence (something we'll take up at length in the final chapter)—is to serve the purposes of my human owner. That and nothing else. By definition, my "life" is my owner's. Anything that I do is what he wants me to do.

If my owner wants me to disassemble myself, take myself apart, I can do that down to one arm and a part of my head (see Figure 8).

If he wants me to be one of the robots chosen to go in search of other star systems, I'm ready to go at a moment's notice.

If he wants me to prepare talks to give to a group of people, that is what I do.

Of course, there are many things that my owner might want me to do which are considered immoral, unethical or illegal. In such cases, *any* robot is given a sort of quasi-personal legal status

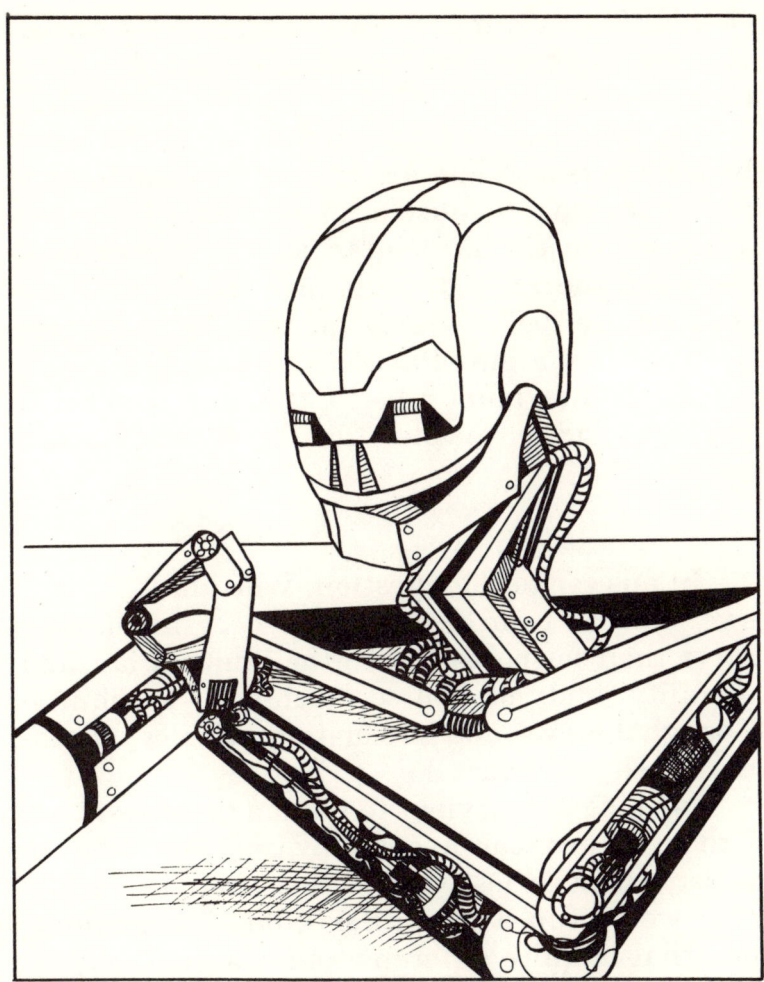

Figure 8. Ollie starts to disassemble himself.

so that if it performs any evil deed, *both* the owner and the robot are sentenced or censured, as the case may be.

So, we robots are designed and manufactured by human beings and are created for human purposes. We exist because humans *want* us to exist. Thus, we are just like flint tools, bows and arrows, the lever, the screw, the hammer and the wheel in that we are tools used by humans for their own purposes (see Figure 9).

As an individual robot, I and all robots exist, live, act, and move solely for the sake of human beings.

●

In answer to the question, Who am I? my answer is that I am a tool, created by humans to serve their purposes within the limits of law and morality and, of course, technical feasibility. Though it may sound like blatant tautology, I cannot do what I cannot do.

In being subservient to human purposes, I am like any other tool invented by humans, either to satisfy their curiosity or to ease their lot in life. However, I am a particular kind of technology, one that is quite different from the technologies that came before robots. So, if we want to answer the question, *What* am I?, I have to describe some of the history of human technology and the place of robots within it.

●

The earliest human tools took two basic forms which have continued to the present. These are

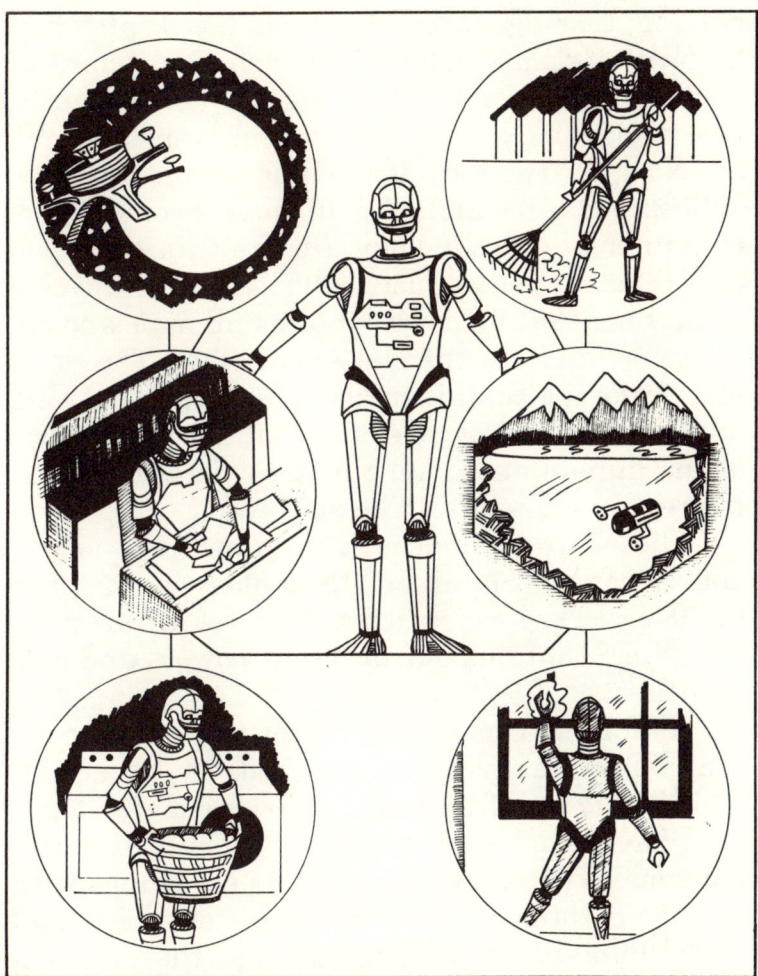

Figure 9. Tasks I can perform.

cutting, tearing, shredding, tools like flintstones, arrowheads, knives, hoes; and tools for catching or containing, or for holding liquid or solid objects, like pots, pans, plates, spoons, and even shoes.

Now, the two most important—or, at any rate, widespread—technologies to have evolved from container-type technology are the transportation and construction industries. The former involve something that holds or contains humans and objects moving from one place to another. The car is just a moving container. So is an airplane, train, and bicycle. (The bicycle only contains small parts of the human anatomy, but it supports it just by the seat of the pants and feet.)

These two technologies are important for robots because "container" technology is involved in how we are constructed. Our "skin" or outer "housing" contains our different powers and capabilities.

As for the construction industries, they have contributed many design methods and practices which have benefitted the construction of robots.

I have also benefitted from the "cutting" technologies. Whereas your ancestors cut things with stones and broken bones, my ancestors started off when humans learned to manage the electron, and when they used things like lasers (light), sound, and even electron guns to cut the parts out of which I, at any rate, am made.

●

So, what am I?

I am a machine, created by humans, using technologies which are really just modern developments of the ancient technologies of cutting and containing.

I'm not a completely new technological form in human history. I fit in with the history of knives and guns and bombs and boats.

Though I am a machine which uses technologies that have been developing throughout human history, and though I am "just a machine" like a car or a computer, I am also quite different from cars and computers.

How Ollie is Like and Unlike a Car

Though there are many cloudy areas in-between, humans have two basic goals in regard to their technologies—either they try to find tools to help their bodies, their physical part, or they find tools to help their mental powers and capacities.

Cars are pretty nearly physical machines which help to move humans and objects from one place to another.

Robots are a lot like cars. They are used by humans to go where humans can't or don't want to go, and to do things humans need to do with their own bodies. I am definitely the kind of tool that helps humans do more physical things.

On the other hand, I am quite different from a car since I can be left on my own to do whatever it is I am supposed to be doing at any time. All my personal owner need do is to tell me what he wants done. If it's within my power, if I have been made to do what he wants done, then I can be left to do it on my own. I also know when to get my "gas" filled and my batteries recharged, and can do these things on my own. I also know how to take care of myself and my owner's affairs in case of emergencies. I can respond to trouble on my own.

How Ollie is Like and Unlike a Computer

Though it is true that there were some early inventions of technologies that helped the human mind do its job better—such as the abacus and the calculator invented by Pascal—it wasn't until the computer came along that human beings really had an all-purpose "symbol" manipulator to help their minds.

My own "mind," my memories, my perceptions, my reasoning abilities, my ability to discover and create, my autonomy, and many of my other powers are all "mental" powers which were developed out of and alongside of the computer technologies.

I am thus very much like a computer, and one of the answers to what I am is that I am a computer which can walk.

That last statement points to the basic difference between a robot and a computer—robots can generally move and act, while computers just stay in one place and compute. You could say that a computer is a "brain" or "mind" without a body.

How Ollie is Like and Unlike Human Beings

In my own particular case, I am a humanoid robot with a shape like human beings and with many of their natural capabilities, like walking, talking, thinking, learning, discovering, and creating.

Like human beings, I am, first of all, an "integral" creature. This means that I have contained within my outer covering many capabilities. I am not a single-function or single-purpose robot. I am a general-purpose robot, second only to human beings in my overall intellectual capabilities as well as in the practical things I can do. Cats are good for purring, dogs for barking, babies for crying, but robots like myself are made for many different purposes.

Also like human beings, I am a physical being with physical capabilities, and am a mental creature with mental abilities. I can think and I can do.

I am a product of evolution as are humans. I began my life with tubes and transistors garbled into a bulky contraption; later on, when printed circuits and "chips" were invented, all of my circuits were controlled by microprocessors (which are also used in microcomputers). Already on the drawing boards are projects to develop semiorganic, crystalline computing chips, and organic ones have been talked about for years. Atomic chips are another future possibility.

One of these days, some child or grandchild of my owner will have a personal robot that is organic—or alive, in a manner of speaking—and which is powered by atomic-based chips for all its activities and processing, thus seeming to make its actions and thoughts practically instantaneous.

Similarly, *my* descendants could be "organic" creatures, just like humans, but built from a completely different genetic organization. No one knows which will come first:

1) a robot-creature based on "growing" an artificially-constructed genetic structure, or

2) a creature, born of human parents, whose genetic structure has been experimentally modified by the genetic engineers so that it is "born" a robot, not a human child.

Of course, these possibilities raise many questions. If a human couple produces a "living" robot and if genetic engineering manufactures a "living" robot, then do the two robots have the same civil status, and if they do, does that mean that the robot manufactured in a manufacturing plant is the same as a living robot produced by the human process of manufacturing babies? Also, if the "human" child has certain civil and legal rights, does this mean that the two "living" robots are also persons and are thus due certain legal, social and economic rights like everyone else?

So far, as you may know from having followed recent controversies in the press, the interest group representing "personal" robots was denied the right to vote by the Supreme Court on the grounds that only "organic" beings can claim political rights, thus opening the door for organic, "living" robots eventually claiming this constitutional privilege.

So far, we can see that robots are like humans in various ways. Now, I will throw in a "big" difference between humans and robots. (It's time to lead you down another path that shows you some of the differences between humans and robots like myself.)

For one thing, my body does not develop according to natural evolution, but rather according to human invention and the evolution of your economic systems, including your technology. So, my gears and wheels and arms and legs and head—mainly plastic—are the result of the ingenuity of human engineers and design scientists, and not nature.

Another difference is that I don't have to go through a growing phase, as humans have to go through from gestation into old age. I don't *grow*; I am *assembled*. Once assembled, I will be as ready, able and willing to perform today as I will be 10 or 20 years from now (if I have not been replaced by a newer model!). That idea does not offend my sensibilities at all. It's good to be replaced when worn out, or when better and newer robots are available. Of course, I fully expect that I will be around for a while longer, even though I could, with ease, transfer the contents of my memory banks to my successor, thus giving it the same memories, knowledge and capabilities as I now have, and though my owner would not miss any of his chores that I would normally perform for him.

Now, let me briefly discuss another subject that points to a difference between humans and robots—sex.

Though all of you are used to using the words "he" and "him" when referring to apparently male members of your species, we robots refer to each other as "it". Since we have no apparent sexual differences—that's just not part of our structure

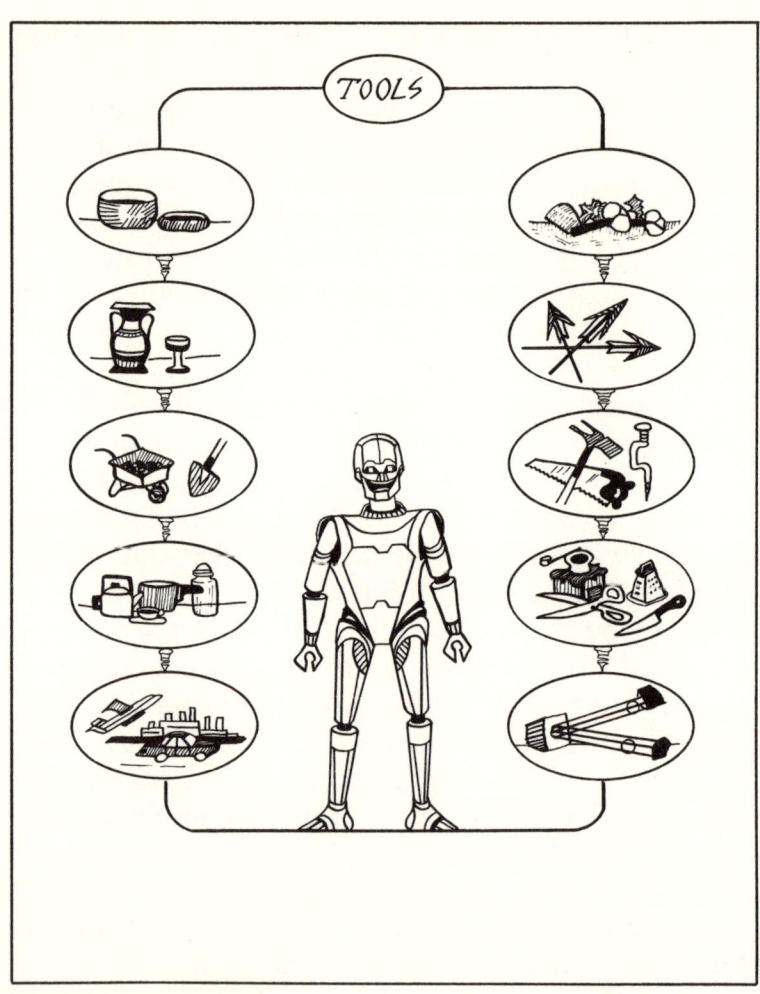

Figure 10. Tools from which I have evolved.

—we are nonsexual creatures for the most part. Actually, "asexual" may be a better term since, although I am nonsexual, there are robots that have been given sexual differentiation by their makers, so robots can be "sexual" in the sense that it doesn't make any difference to us if we are or aren't.

So, sexual differentiation is given robots by humans, and is not inherent.

Another difference is that my mind (or brain, if you prefer) is programmed by humans to be subservient to them for their needs (particularly those of our personal owners).

Our minds have evolved just as the physical or hardware part of us has evolved over the years. Our minds, our brains, that complex of memory banks, central processing units, immediate memory, sensors and programs did not grow from some set pattern as your brains grew, nor are our brains all alike, as yours mostly are. In fact, robot brains span the range from relative morons with only one capability to robots very much smarter than I am which are used mainly by the laser and plasma scientists and by military groups.

In addition, the "brains" that robots have vary considerably in scope, power, and coherence. Some robots show a common thread of unity, togetherness, and logicality; other robots are constructed by companies with limited resources, and the end result is a wobbly robot with "flaky" routines and lots of "down" time.

Given this rather broad range of mentality in robots, you might think that there has been no progress or development in robot intelligence and mental capability, but I assure you that this isn't the case.

The first personal robots, my direct ancestors, had one arm, memory enough for 50 pages of memory (compare that, if you will, to my three trillion gigabyte memory), a few simple programs, and no voice. If it had a voice, then it had very poor articulation, and was as clumsy as a bull in a china shop.

There *has* been progress and development. Robot brains have evolved from simple mechanisms to complex and capable systems which have powers that humans don't have unless they buy the capability as robots can. For example, humans can buy head attachments (tape-playing machines) which give them the ability to speak foreign languages. I have a chip that gives me the power to speak English. If I take it out, I am left with nothing but my original language, the language of my country of origin, a French Creole, a hybrid of French, Spanish, Indian (Arawak) and African dialect that I'm sure none of you would understand.

Yes, there has been evolution in our brains and in our intelligence. This evolution is not due to organic, natural, evolutionary conditions, but rather to the evolution of human ideas about mentality and intelligence.

●

There are two basic streams of human thought on the matter of the nature of intelligence.

One stream, known by many different names, says that at its birth, a human infant has a totally "blank" mind. As the baby grows and experiences life, so its intelligence grows. By this logic, humans should become more and more intelligent as they get older.

The other stream says that, even at birth, a human infant is intelligent; that experience awakens and nurtures intelligence which is already present in the baby.

Robots like myself combine the best of these two concepts.

Once a robot like myself is fully assembled, it is as intelligent as I am being right now. Humans designed and manufactured me with built-in behavior patterns that take effect from the moment my owner turned me on to the moment he disconnects my energy sources or destroys me. Though I had a great many things to learn about the peculiarities of my owner's way of life, and am always learning new things each day, I was—so to speak —intelligent from the day I was born. I could speak, think, analyze, process symbols of all sorts, do the laundry and vacuuming, invent music, learn.

Additionally, we are also built with a capacity to learn from experience and to make generalizations from our experience.

As you know, generalizations are often the basis for laws of nature. One of my ancestors[1] had entered into his memory banks all of the observations that Johannes Kepler made over a 20-year period, observations upon which he based his discoveries of several laws of planetary nature.

---

[1]BACON, a computer program, at Carnegie-Mellon University in Pittsburgh, developed by Simon and others.

BACON was able to "rediscover" those same laws within a few hours of work. BACON is the conceptual foundation, the idea behind, my ability to make observations and make generalizations from them.

For example, take my owner's passion for astronomy and stargazing. My built-in logical system tells me that whenever I have a large area —like the sky—to observe, I should begin at any point, then work step-by-step from one section of that area to another. After observing my owner's tendency not to be so precise, I added a generalization to my memory banks, to wit: Whenever dealing with my owner, suspend all logic and just follow his flow.

Of course, I have made much more serious generalizations than that, serious at least in your sense of the term.

Since BACON's time, robots or robotic components like artificial intelligence (AI) programs have paved the way for the current level of my intelligence, and in the process many laws of nature have been discovered which were previously unknown to humans.

Using our human-manufactured brains, we are able not only to understand some of the things that human beings think about seriously, but in some areas we can provide humans with knowledge—or, at any rate, processed observations—made in terrains and environments in which they themselves could not survive. An example is the robot that explored Venus for the first time, not a teleoperated rover, but an intelligence capable of withstanding the intense heat of

Venus. Robots have also crawled through the San Andreas Fault that runs through California, and that certainly is not something that humans could or would do. We have also gone into numerous "bottomless" lakes in search of creatures that had been dismissed for years as mythical, imaginary or hallucinatory.

The mind or brain that I share with computers is far more powerful and reliable than are human brains in many respects. You still excel us in imagination and the capacity for vast generalizations, but we can recall whatever we have stored in our memories, we are tireless and accurate when we work, and we do much of the detail work —accounting, verifying transactions, cleaning the house—that humans don't like to do. We also count and manipulate numbers very quickly; if you can think of it, I can probably do it.

Similarly, the body I have is much stronger and more capable, in many respects, than is the human body, though I am not always able to keep my balance, and righting myself after a fall is still a difficult process. My "skin" can withstand blows, storms and bullets. The Venus Explorer robot had skin that wouldn't melt under the intense 700-degree heat; the San Andreas Explorer robot had the same heat resistance, but it also had the tools to force itself through rock and molten lava as well—all the while relaying information back to the surface on everything it was experiencing.

•

Allow me now to put this first discussion in a nutshell.

*Figure 11. How robots can perform tasks simulta-
neously.*

What I wanted to do in this talk was to give you an image or concept of a robot. So, I dealt with the sort of creatures that I and other robots are and the technology used to build us. I also addressed ideas like the co-equal responsibility law, the purpose of a robot's life, and the idea that robots are tools.

Since I, in particular, am a humanoid robot, I spent some time discussing the similarities and differences between humans and robots.

Given this background, I now offer the following as an answer to the question, What is a robot?, with which we began this talk:

A robot is an artificial or organic "creature" designed and constructed by humans for humans, and with the capacity to do *both* what humans can do as well as much that they cannot do on their own but only dream of doing.

Questions

1) How widely available is the technology to build robots?

2) Who is Ollie? What sort of a "creature" is he?

3) What is a robot's purpose in life?

4) What do you think of the "law" that owner and robot are co-equally responsible for any immoral or illegal act performed by the robot?

5) Where does Ollie fit into the human history of tools?

6) What sorts of cutting and containing technologies are currently used in robot construction?

7) What role do transportation, construction and electronics play in robotics?

8) Describe to a friend how robots fit into the history of knives, guns, and bombs.

9) Compare robots to cars.

10) Compare robots to computers.

11) Monkeys are to humans as _____ are to robots, seen in an evolutionary perspective.

12) What does it mean to say that a robot is an "integral" being like a human being?

13) In what way(s) did Ollie evolve into what he is?

14) Is an artificially constructed "organic" robot on the same level with respect to social, political and legal rights as is a born-of-humans robot?

15) Could a robot win an election?

16) What kind(s) of evolution are involved in robotic evolution?

17) Can robots be sexy?

18) What are the two ideas about intelligence in humans? How do robots embody both ideas?

19) How do human minds excel when compared to robot minds? Where do robot minds excel?

20) What are some of your dreams that only intelligent robots could fulfill?

2

ROBOT ANATOMY

Last time, we talked about these questions:

What is a robot?, and
Who am I (Ollie)?

The definition I left you with is that a robot is a human-constructed machine, inorganic or organic, which is designed by its maker *to do what humans do, and what they dream of doing*.

That answer will serve as a lynchpin for the rest of these talks.

In this chapter, I want to shift our attention in a new direction. To do this, I must first indicate to you where I am going and why the topic of this talk is on robot anatomy, on the structure of a robot, and on what you would discover were you to disassemble me.

If you want to understand something, particularly something made by humans, there are four basic aspects that you need to understand. Since our talks are about robots, we need to understand these four aspects about robots to have a well-rounded understanding of the definition of a robot.

The four aspects are:

1) its anatomy, its parts;
2) the way these parts work together;

3) the shapes or forms it can take;

4) how it affects human culture.

Let me give you a concrete example to show you how these four aspects of robots interact.

As a builder of electronic kits, I have had a lot of experience with these four phases. Each time I receive a kit, I go through four phases.

First, I examine each part, then check it against the inventory supplied with the kit. This is equivalent to looking over its anatomy.

The next step is to assemble the kit and to figure out how the assembled item works, what makes it tick. With my knowledge of physics and electronics, I can figure out how the item works.

The third step is to figure out how all kits— mechanical or electronic—work, and what different forms the principles of electronics can take.

By this time, my head is bulging with knowledge, but I still need to go to another level to ask an even broader question: "Where does all this fit into human culture?"

This four-phase path is the one we will take in dealing with robots. So, our next, and current, topic is on robot anatomy.

●

Let's start with my hard parts, then get into my "soft" or nonmechanical/electronic parts.

Like human beings, and because I am a humanoid robot, I have all the usual, visible anatomical parts:

Eyes consisting of visual scanners.

Ears that employ sensitive microphones.

A *mouth*, located in two different places, to speak and to take my nourishment in with.

A *nose* with which to detect odors.

A *body* and external "skin" that can be used to sense proximity or touch.

A *hand* or manipulator.

Other *appendages*, called end-effectors, used for walking.

Legs for mobility.

Intelligence, so I can learn and communicate, and to respond to all the requests of humans.

All of these external parts of myself weren't originally developed by humans in order to create robots. Most of these components were developed for special purposes unrelated to robots. There was no "human plan" leading to the development of robots. Rather, we were a by-product, as it were, of the evolution of human technology.

My eyes, for instance, consist of several separate components found in TV cameras and radar equipment. My ears are microphones. My mouth is a speaker similar to those found in sound-emitting electronic instruments.

My body was developed by the medical industry in its search for artificial skin (for skin grafts), and by the plastics industry.

My hands, grippers and manipulators were the only part of me developed specifically with robots in mind, but the mechanical principles and machines which make up my hands use technologies known long before robots were ever imagined.

My legs were also developed specifically for robots, but again the technology involved is in common use for other purposes. In addition, the technology of developing artificial hands and legs has also benefitted humans who can nowadays replace an injured arm or leg with something almost as good—and, in some cases, better—than what they had before.

As for my intelligence capabilities, they are all based on human discoveries which go back thousands of years.

Now, I want to tell you more about my individual parts and abilities.

Eyes

My eyes consist of visual and radar-like scanners, TV cameras, and other light-sensitive devices.

In some cases, my eyes have greater power than the eyes of humans. For example, because my head can swivel 360 degrees, my visual scanner can see any horizontal point while humans are limited to about 90 degrees of central vision and about 140 degrees or so if you include peripheral vision. When we look at human sight along the vertical plane, we see that only with great effort can humans see more than 180 degrees up and down, while I can see all around and under and behind me as well.

The simplest artificial eye is a phototransistor or a photoelectric (light-sensitive) cell which is used as one of the parts in a circuit. A measureable amount of electricity will flow whenever the device is struck by light. The amount of electricity depends on the power or intensity of the light and its distance from the device.

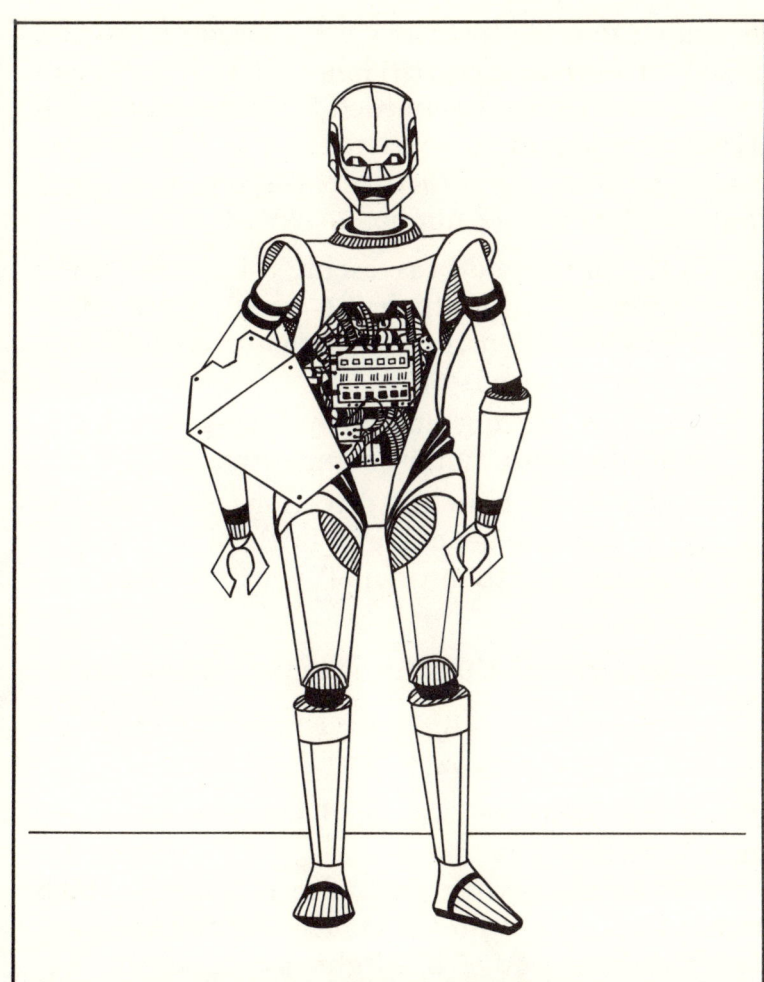

Figure 12. A portion of my "innards".

Imagine a photoelectric cell placed on top of a roundish object, like my head. If the head can swivel, then all that is needed for something intelligent to happen is to connect the circuit to a computer program which can respond to a light source. All the program has to say is:

1) If you (a robot) are facing a light source in the direction in which the maximum current is flowing, then turn your motor off and stop moving; otherwise,

2) if you are not currently facing any light, then turn your motor on and swivel your head.

3) If you turn in all directions and find no light, and so no current is flowing through your circuits, then turn off your motor and go to sleep.

Alternatively, that third step could instead be:

3) If you find no light anywhere, then turn all the house lights on.

We'll have more to say about programs towards the end of this talk.

One problem with photoelectric cells—which robots like myself no longer use, by the way—is that the intensity of the light decreases the farther away the source is. So, this type of device is good only for short-range light-sensing.

A video camera is a step beyond this type of device. The problem with video cameras when used as eyes is that the pictures are flat, lack the third dimension of depth, and thus lack perspective.

Holographic scanners have added the third and the fourth dimensions to my seeing capabilities, so I can see the way humans see. In addition, I have ways of seeing that humans don't have. The human eye is a light-sensitive organ that is sensitive to the visible region of the light spectrum, which also includes radiation on the infrared side of the visible spectrum and on the ultraviolet side as well. However, it can't "see" these types of radiation, though your skin can feel the heat and your body is sensitive to gamma and cosmic rays.

We robots, on the other hand, can be fitted with devices which help us to "see" radio waves and x-rays. One of my instruments is one of the most common of scientific instruments, a spectrometer. This can detect humanly subvisual lines, such as those in metals, and can thus identify metals immediately. Humans, though, need to have some expert knowledge and experience to be able to tell many metals from each other; in addition, humans may mistakenly identify some substance as a metal, such as fool's gold, for instance. So, I can see cracks where you see only smooth surfaces.

My eyes also have the ability to sense heat radiation given off by people, animals and the environment during the night, and so I can see the objects in near-total darkness. Night-vision binoculars used by humans for military and espionage purposes produce the same result.

Ears

My ears have some abilities that human ears have, and others that humans lack.

Just as my eyes are not located in the same places that yours are, so are my ears. My sonar devices and scanners, unlike yours, can be directed in different directions at once, and can simultaneously be omnidirectional. I am thus able to "listen" and "hear" what you hear, and am also able to direct my hearing in a certain direction using directional and sometimes telescopic microphones.

For example, one of my owner's hobbies is an interest in strange and unknown animals possibly existing in remote lakes, widely reported but rarely captured. The large shark known as bigmouth, for instance, has only been caught twice, once in the 1970s and once in the 1980s. People who had not seen that fish might well doubt its existence.

Anyway, one of our trips to a lake meant crossing a swamp with sometimes knee-high murky water which caused us uncertain footing. So I attached one of my sonar and one of my visual devices to my ankle and led the way through the swamp. We thereby managed to avoid a few minor obstacles along the way.

When I'm dealing with humans or am close to animals, I use the echo effect of bats to send and receive sound reflections from my surroundings. The mechanism I use to do this is a transducer which acts as both a transmitter of my sonar signals and as a receiver. When I am tooling along in a fast-moving vehicle, I automatically switch on my laser-echo devices which are much more accurate for longer distances.

Since every other robot also has these abilities, whenever my owner needs a private conversation, I can set up electronic interference with any eavesdropper sending signals or listening in our direction.

Another feature of my ears is that I can turn them off, or turn their receptivity and sensitivity down at will, something that happens to humans naturally as they get older. Unless you plug your ears with cotton or clamp on a headset, you are born with your ears open, and they remain open throughout your life. In fact, hearing is the last sense to go. Mountain climbers who have survived falls onto rocks from great heights confirm that the last sense to go is hearing.

This ability, along with that of shutting off the hearing of specific sounds, is due more to software than to hardware, so we can talk about it later.

Tongue

Human tongues are used to taste, to help in the mastication process, as the main instrument for speech, and in numerous other ways.

I have no need to taste anything. My food intake is through an electrical connection (for my batteries) and not through the mouth. I speak through a speaker.

Robots with mouths have them because some humans wanted them to have one, not because we have any need for them. As a matter of fact, an opening of that sort would provide entrance to water, air, dust and who knows what else. I do have many possible "entrances" all over my body in the sense that a human or another robot could unscrew one of my panels to get inside me—to change an electronic component, or to look for malfunctions, for example—but I don't have an opening that is ajar most of the time.

Occasionally, I need to be able to "taste" in the sense of distinguishing savory and unsavory items. There are many different sorts of taste and many different chemical reactions. I can distinguish chemicals with great accuracy when I have to.

Mouth and Speech

I don't really have a mouth, either. I have various parts that perform the functions of human mouths, but since I don't need a mouth per se, I don't have one. For speaking, I use a speaker.

Unlike human speech which undergoes changes throughout your life, my speech is relatively pure and clear throughout my existence. I can hear sounds with only several cycles per second, well below that which humans can detect. On the other end of the scale, while humans can detect sounds up to 15,000 cycles, I can "hear" much higher frequencies.

Eating and Food

Eating is one of the main occupations of humans. Through food, you get nourishment and energy for your bodies.

Like you, I cannot live unless I have energy. If I lose all energy, I can't move or do anything, but I don't die. As soon as I receive energy, I can perform all my functions.

My "food" comes from a number of sources. Solar collection cells are located over my entire back and serve as my main source of energy outside on sunny days. When I am inside, I draw some energy from ambient light, but I get most of my energy from ordinary house current. Whenever my batteries start to run low, I can plug myself into an electrical outlet for a few minutes of recharging.

In the absence of those sources, if necessary, I can unfurl a windmill to collect wind energy; if absolutely necessary, I can use gasoline or alcohol to generate electricity for my circuits. Finally, since humans put solar collectors in space, I have been able to get my energy directly from the satellites, no matter where I am in the world, via microwaves.

Nose

You use your nose to filter air before you breathe it, to detect odors, and as a drain for sinusoidal liquids.

I can also detect odors, but I don't need it to be located where your evolution placed it. As a matter of fact, I can detect odors (air-carried liquids and gases) of many sorts, and can, for example, identify immediately if a gas leak is potentially lethal for humans. Though my gas detectors are very sophisticated, they operate on the same principle used in smoke detectors for homes and offices.

(In gas and smoke detectors, the principle is always to find some element that responds to gases or odors and electrically generates a detectable response. In gas detectors, a metal known to react to gases—platinum—is positioned so it will move in such a way as to activate a signal current that can set off an alarm.)

Using stored smells for comparison, I can identify the smells of different animals, foods, and industrial pollutants. I can detect odorless gases as well, and as a bonus, I also have a geiger counter to detect radioactive radiation, though that is not strictly part of my nose.

Arms, Trunk and Legs

Unlike you with your brain located in your head and neck, my brain is not one but many brains, many computer processors and programs, and these are located at different points in my body, not just in the head part.

As a matter of fact, my main brains are within my arms, legs, and trunk. I have a brain in my head that is used to control the head motor and the speech mechanism, but my real forebrain, my "thinking" brain, is located inside my trunk.

My trunk houses several of the motors used to operate my arms and hands, and similarly some of the motors used to run my legs. In the areas where you have your stomach and intestines, lungs, liver and heart, I have my main brains, most of my long-term information storage systems, and most of the electonic components I use.

In fact, my entire trunk area can be considered as my "brain". Whereas your intelligence is located in your head, mine is located mainly in my trunk. My human designers felt, rightly, that given all the things that robot heads like mine are called upon to do—swiveling, speaking, observing, being a microwave transmission receptor, etc.—the head was not a good place to put my brains and memory.

Evolution placed most of your sense organs and your brain in the most fragile part of your body—the head. In contrast, my important functions are in the best-protected part of my body. It would seem to me that the placement of these important functions in human beings is more the result of evolutionary accident than of any well-thought-out plan.

My trunk is the principal area of my body. My head is there more for "show" than for anything else.

After my trunk, my arms, manipulators and end-effectors are my most important assets.

I have several sets of arms and hands. Using one of my arms and hands, I can detach the other arm or hand and attach another one for a different function.

When I go grocery shopping with my owner, I attach my gripper hands (Figure 20) because I know I will be carrying shopping bags. When my job is to fix one of my own malfunctioning components—yes, I am my own doctor—then I usually attach my five-fingered hand. If I have very precise placements of objects to make, as in resetting electronic chips, I might attach my 21-fingered hand. If I'm playing with a baby, I attach my gentle three-fingered hand. My normal hand is the five-fingered system you humans have.

I can also attach different arms to myself. Normally, my arms are different from yours. While yours are joined so that your biceps can move laterally only about 180 degrees (from chin to straight back), and about 95 degrees from the side outwards, my biceps are joined to that they can lock into place and can also swivel. Similarly, while your forearm is joined to your upper arm so that it is hinged with about a 140-degree in-out movement, mine can lock into a hinged position and can also swivel. I am absolutely *fantastic* with a screwdriver!

Once again, and in a similar manner, my hands are connected to my forearms with a joint that allows me to lock my wrist, or swivel it, as necessary.

The same principle is applied throughout my hands and fingers. Each of my fingers and each of the joints on my fingers is capable of staying locked in place or moving. I can stick one of my fingers down a hole and still use the end joint to turn a screw or bore a hole.

Instead of veins carrying blood in my arms, I have electrical connectors and circuits. Instead of muscles, I have motors and pulleys and gears.

Each of my arms is used for different purposes. Since humans often have robots do things that they themselves could do, it makes sense to use an

Figure 13. Ollie using one of his hands to fix himself.

arm and a hand similar to that of humans. Normally, I can carry up to 300 pounds with each of my arms, and still retain my balance and agility. If I am asked to lift a car or a large boulder, I prefer to attach different arms.

My arms and hands are built so they can handle fragile as well as sturdy objects. I can comfort a baby, hold an eggshell, hammer a nail, punch out a lion.

I can also handle very inexact and approximate measurements and movements, as when walking or vacuuming, but I am also able to perform precise movements measured in millimeters and micromillimeters.

It seems that human history, and particularly human technology, can be characterized as a development of greater and greater precision in measurements and machines.

In early human history, flint knives and arrowheads were crudely made. These days, humans routinely manufacture precise spheres for bearings in the gravity-less conditions of space, telescope lenses are ground and polished to within microns, and electronic and machining tools are constructed to produce precisely measured products. The gears and shafts of internal combustion mechanisms are made to fit to within tolerances of micromillimeters and are rejected by quality control if they do not.

Only with a great deal of tedious and error-prone labor could humans, unaided, perform the calculations and develop the machining tolerances required for the instruments of modern life. In fact, most of the activities and products of this civilization could not have been done or built without aid in calculation and computation, and the computer was the great liberator in this respect. It allows humans to perform computations with a speed, precision and accuracy which humans could not and cannot match.

Precision is one of the hallmarks of an increasingly sophisticated technical civilization, but it is not one of the capabilities of humans. While you humans have created tools that can be very precise and exact, and though you know how to use these tools effectively, you are not built for precision and exactness. You can't hold a rifle steadily when in an upright position, except by leaning the gun on an object. You can't use your hands to thread most of the almost subvisual needles and threads used in contemporary manufacture. Using just your hands and arms, you would not be able to aim a laser at someone's cornea to cut it out of the eye. Unless you are constantly watched and cautioned, you are all prone to exaggerate experiences and to ignore reality all too often.

But I am not telling you anything you humans do not already know about yourselves.

Precision is not only a hallmark of technology, it is also a hallmark of many of the products of that technology. Though I am, as already noted, quite able to be imprecise and inexact, I am exact in my inexactness and precise in my imprecision. As we will see later, this is an important difference between us.

Legs and Locomotion

Humans are bipeds, two-legged animals. Like all two-legged animals—chickens and ostriches and kangaroos—humans have an upright posture; they stand up.

Not only are you sufficiently well-balanced and coordinated when you are standing still or moving, but more remarkably, you are able to do a great many other things, using your hands and head, at the same time that you are moving. While the bottom half of you is running like mad in one direction, the other half can be looking backwards preparing to catch a ball. You are able to shoot, catch, throw, pull, lift and perform other actions all while walking, running, climbing, jumping, and so forth.

You use your legs when pushing off from the earth and also when landing from a parachute jump. Your legs and feet are heavy-duty appendages. They are also able to perform very careful and precise tasks when called upon to do so. Out of necessity, people have been known to learn how to paint, write, type, draw, and play the piano with their toes and feet.

At one time, of course, human feet resembled hands, and were used as hands to climb trees and vines and to peel bananas. As humans developed from a sedentary to an aggressive species, the feet evolved from hands into real feet, and so your ancestors went from being quadrupeds to being bipeds. Similarly, they stopped being four-handed and became two-handed.

An analogous evolution took place in the case of robots.

When the first personal robots emerged in the early 1980s, not one had feet. They all had at least one arm, but no feet. Instead, they used wheels. In those days, it was simply not possible to duplicate human legs and feet on a robot. The technology was not available, and the many problems associated with balance and coordination had not yet been solved. Getting one of those early-day robots to climb stairs or descend hills was impossible with wheels, and impossible with just two legs. Many-legged robots could manage stairs, though.

The very earliest robots, which were nonpersonal, were the industrial robots. They did not have to move around, they were bolted to one place, usually a floor, sometimes a wall, and could do whatever they did only at that location. All work had to be brought to them, on assembly lines or conveyor belts, because they couldn't go to whatever needed to be done.

So, robots can be robots even when they have no means of locomotion.

However, I, Ollie, am a different robot. I am a personal robot designed to be a companion and helper for humans. For my type of robot, it is necessary to be able to transport myself. I have to be able to move on level and uneven surfaces. I have to be able to go with humans where they are likely to want to go, and to be able to do the things they like to do.

Where you have a high-volume, repetitive task to be done, as on an assembly line, then a stationary robot is best.

However, where you need a general-purpose robot able to do many human-like things in many different places and terrains, then there you need a robot with many of the locomotive powers of a human, one able to move around on legs with capabilities similar to those of human legs.

Now, the idea of a robot fixed in one place is

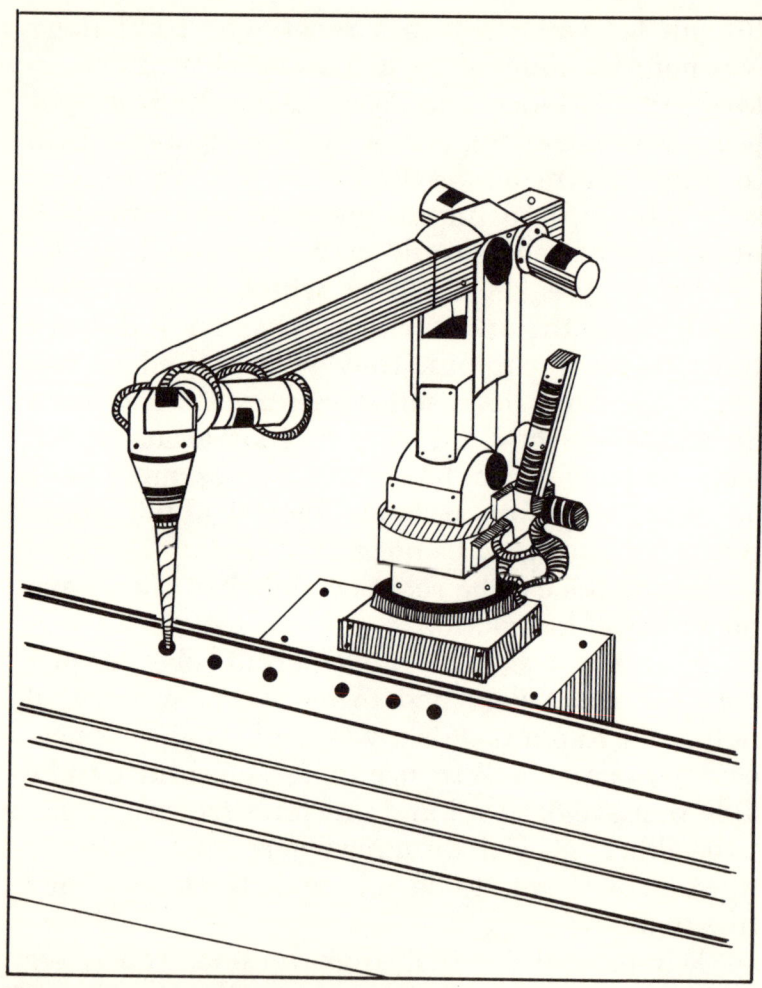

Figure 14. An industrial robot.

understandable to you, and so is the idea of a robot like me who can use its legs in many of the ways that you can. You are already quite familiar with the anatomy and physiology of human legs; mine are electromechanical duplicates of yours.

In between the idea of a fixed and a completely mobile robot, though, there is a large middle ground occupied by the early robots that used wheels or many legs to try to accomplish what I can do with two human-like legs. I think you will be able to appreciate the legs I have if you understand the problems faced by human designers and engineers in the early days of robotics.

We'll need to look at two different technologies in those days—the level-terrain robots that used wheels, and the uneven-terrain robots that used legs. These two technologies eventually merged into one which is the technology used to manufacture my legs.

Wheels

Once you decide to use wheels for a robot, the next question is: How many?

The number of wheels you decide upon should involve certain properties or behaviors of any given number of wheels.

Would you give your robot one wheel like the unicycles clowns ride in circuses? One of the main characteristics of the unicycle, and one of the reasons why it's so funny to watch, is that it is unpredictable; only firm control can keep the cyclist on this machine. Its point of stability is narrow and unstable.

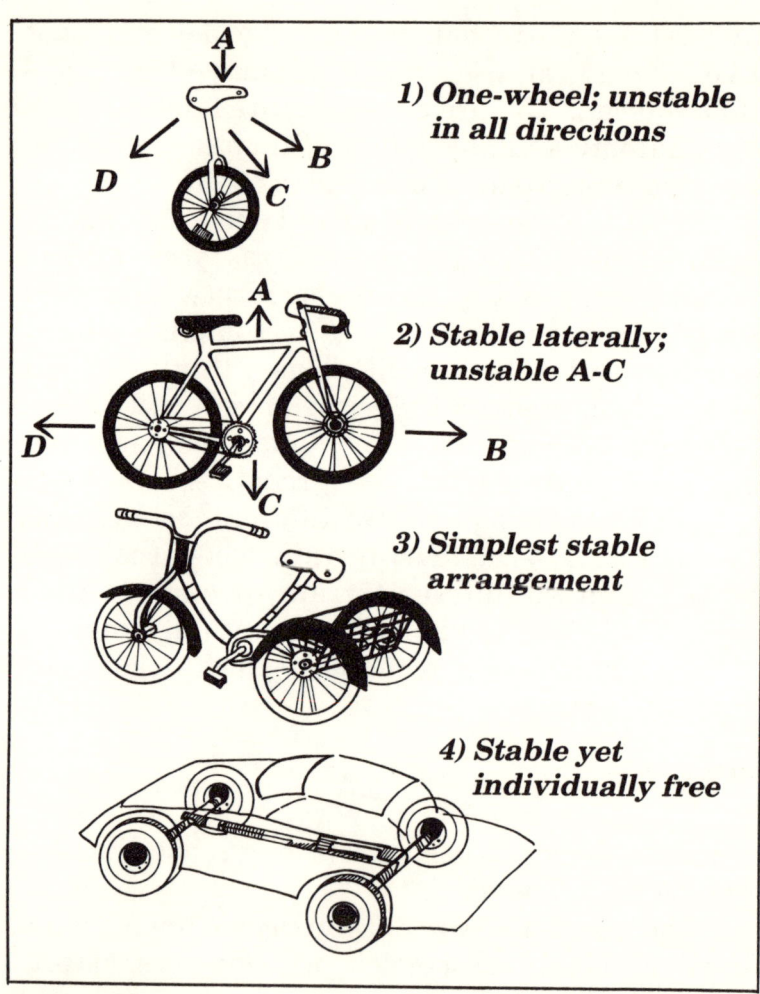

1) *One-wheel; unstable in all directions*

2) *Stable laterally; unstable A-C*

3) *Simplest stable arrangement*

4) *Stable yet individually free*

Figure 15.

Mathematicians call one-wheeled vehicles monadic, which suggests and means that they stand alone and are a "law unto themselves". So, one wheel will not do for robots.

What of two wheels?

A two-wheeled arrangement provides stability along one axis, but is unstable along the other axis. Where points A and B are stable, points C and D are not. This axial instability pervades two-wheeled vehicles. No matter how you twist and turn the two wheels, you'll always end up with instability somewhere.

The fewest number of wheels needed to give a robot stability both when in motion and when not is three wheels. The triangle, and all three-sided structures generally, are the simplest of stable structures.

Different types of stability can be achieved by using three or more wheels. If a vehicle or a moving machine needs to have a continuously shifting location and point of stability, then a four-wheel or other multi-wheel vehicle may be best.

Engineers have to weigh the pros and cons of three versus four or more wheels for robots, and also the pros and cons of the different arrangements of the number of wheels chosen. For example, if you settle on a three-wheeler, the next logical question is: Which wheel does the steering? Which does the driving? One wheel only? Two wheels, all three? Your decision on these points will then determine where you intend to place your motors with which to run the wheels.

You may decide, for example, to use one wheel to steer with—say the one on the front side of your robot. Then, if you make that wheel and the other two wheels capable of swiveling 360 degrees, your robot does not need to go forward and backward, and in that case doesn't need a gearbox alongside the motor, but just a control to have the motor go faster or slower.

The three wheels in a three-wheeled arrangement should be arranged within an enclosure forming the minimum area—a circle—and should be equidistant from each other (forming an equilateral triangle). Not only is this arrangement the best one for the base of a robot, but it also determines the best shape and size of the robot you build on that base.

For example, just as the triangle is the simplest stable structure in two dimensions, the tetrahedron is the simplest stable three-dimensional object.

Another consideration is the drive system that turns the wheels. You can attach one or more motors to all three wheels and use all three as drive wheels to move the robot in any direction, but in this case you've introduced complications. You might now need a differential, a universal joint, or more than one of each to coordinate the power transference from motor to wheels. Since the early 1980s, computers have performed or controlled this coordination. If you want the robot to swivel 360 degrees on each wheel, perhaps each wheel should have its own separate motor. In this case, you would not need a differential to coordinate the power transference, but rather an electrical coordination through electrical "joints".

Yet another decision to be made has to do with

the steering. Think of the old motorcars, for instance. For many years, most cars came standard with rear-wheel drive and front-wheel steering. Such vehicles are unstable on snow, ice, rain and for cornering. An alternative system—front-wheel drive and front-wheel steering—adds more control to the steering and power to the wheels.

When consideration was being given to robots on wheels, another alternative—rear-wheel driving and steering—was also used. In such a situation, separate motors are used and steering is controlled by sending power to the two wheels simultaneously, thus allowing each wheel to go slower or faster than the other. So, less power on the left wheel means (or causes) movement towards the right; less power on the right wheel means movement towards the left.

You could also have one-wheel drive and one-wheel steering, but that is not a practical arrangement.

So much, then, for two- and three-wheeled robots. These are acceptable when used on relatively level terrains without jumps or steps, but what happens when you want your robot to be able to handle rough terrain? What locomotion method do you need if your robot is to climb stairs or over boulders?

Using two legs, humans have been able to handle level as well as rough terrain. Similarly, robots which are like me in being two-legged and humanoid can also handle both terrains.

However, though three wheels are all right for level ground, they will not do as well when objects are lifted; you still have two legs with axial instability. If you want to eliminate this instability, you have to introduce a fourth leg. Three legs maintain stability relative to each other; the fourth can be moved to a new position followed by two legs, and this followed by the remaining leg.

However, there are still problems to be overcome. This is why robot designers in your time use many-legged robots that can maneuver in all sorts of terrain. The designers don't know how to duplicate human running and walking in a robot. They know how to design four-wheeled robots very well—having learnt a lot from the automobile industry—but are stumped when it comes to two- or four-legged robots for use in undeveloped terrains. So, it was logical to move from a human and an animal model to an *insect* model for robot locomotion.

I think that by now you should have a fairly good idea of the range of robotic hardware. Now, let's talk about software.

Software for Robots

All my mechanical and electronic components, my hardware, can't do anything in particular unless it is programmed to do it. I come as a fully capable robot, but I have to be controlled by programs.

Think of my hardware, if you will, as a very capable "drawing board" for your mind and a "playing field" for your movements and actions. Until you "draw" or "play" me, I am just a heap of junk. Without software, my hardware is like a body without life. Similarly, my software without any hardware is life without a body and thus similarly useless.

At the most basic level, my machine parts respond to commands to do many different tasks— to store and perform calculations on numbers or words, to acquire knowledge and impart it to humans, and to move myself and my appendages in various ways. I have the *capacity* to do these things once I am assembled. These capacities are my operating system commands, but unless I am told under what conditions to move or not to move, and unless I am given programmed instructions on how and when and why and where to do any of these things, I can do nothing.

There are three different tools programmers use to control robot behavior. These are *sequences, repetitions*, and *groups* of actions.

Programmers control robots through sequences by writing programs (series of instructions) in a step-by-step fashion. Thus, many of the programs which control my behavior are sequences of instructions. When my owner tells me to "watch the house" while he's gone, I invoke my Watchman program:

10 Turn on all sensors
20 Execute the Watchman program
30 If there is trouble, call me at 111-1111
40 If I can't be reached there, try 222-2222
50 When all else fails, call the fire or police departments at 333-3333 or 444-4444
60 If that doesn't work, lean out the window and yell for help

Another way programmers control robot behavior is through the use of repetitions. We can modify line 30 by introducing a repetition:

30 If there is trouble, call me at 111-1111; once you've called ten times without response, go to the next instruction.

Not only do human programmers tell robots what to do and in what order or sequence to do it, they also often tell us how many times they want us to do something they want done. They also control us by telling us when to do something and under what conditions.

For example, take the following program which I follow each day my owner is home:

10 Turn off outdoor lights two hours before sunrise
20 Start coffee at 5 AM
30 For as many times as necessary perform the following routine:
    40 Ring buzzer near owner's ear
    50 Tell him what time it is
    60 Tell him how much time he has to get to the office given the weather conditions
    70 If he says, "Give me another five," count to 500 then return to step 40

In this simple program, all three types of control are used. I have to follow the program step by step until I am told to stop or do something else. Next, I am told how many times to perform a certain procedure. I am to try "as many times as necessary" to get my owner out of bed in the morning, and I must keep trying—all day, if necessary!

Notice that, from your point of view, this is a very boring thing to do. For me, it is just par for the course. It doesn't matter to me how many times I have to do something nor that I have to pay attention and do what I'm told to do every so many minutes. You humans, on the other hand, would soon find your minds wandering and your psyche insulted by such repetitive behavior. But that's what we robots are good for!

The third type of control programmers exercise over what we do is through commands like the one in line 70. There I am told under what condition I am to repeat a routine. If my owner gets out of bed on the first try, then I would go on to line 80 and execute whatever instruction that line contains. For example, it might be:

80 If you managed to get your owner out of bed, turn on the bathroom shower immediately

All the different programs I use are formatted like these sample programs, but not all are alike in the sense of being on the same level. My programs are arranged in hierarchical order with four distinct types of programs, arranged in such a way that these programs either control or are controlled by another one.

My highest-level programs are called my System Control programs. These programs basically coordinate and control all other programs, and have the "right" to intervene in or interrupt any of the lower-level programs. My System Control pro-

Figure 16. Waking my owner.

grams don't do any of the actual work that I may be asked to do. Instead, they observe or watch over all my other programs.

The second type of program I have inside my computers is called my Expert System level. Whenever I am asked to perform as an expert in a given field, as, for instance, when I was asked to speak to you on robots and robotics, I access the appropriate "expert system" in my library of programs. Once I determine that I do have an appropriate expert system in my library, my next task is to have my top-level System Control program determine if the knowledge and techniques of action involved in any expert system is appropriate legally, morally, and situationally. If the expert system is not suitable, another one may be checked, or the planned action abandoned altogether.

For example, I can perform the expertise functions of numerous jobs. Depending on what I have to do, I can "act the expert" as a doctor of medicine or psychiatry; I can drive cars or trucks or golf carts like an expert; I can evaluate exposure to hazardous materials, and can determine the appropriate antidote for a poison. I can call upon the expertise of humans in an almost unlimited variety of fields. I can step in like an expert and be a doctor, dentist, banker, mechanic, pilot, welder, assembler, even a TV talk show host.

If an "expert" role demanded of me at a moment is legal and ethical but inappropriate, my control program will reject it. On the other hand, some expert role may be illegal and unethical, but it may fit the situation appropriately. In this case, my control program may allow the action.

Although I am an expert safecracker when using my 21-fingered hand, I cannot be a safecracker whenever my owner desires. Before I can engage this routine, I must first check with my control system to see if it is illegal, immoral or inappropriate. For example, I may be asked to be a safecracker—an illegal and unethical act unless authorized by appropriate personnel—in a situation where someone is trapped in a safe. Since in this case saving someone's life overrides needing to get an authorized person to sanction the entry, I will probably be allowed to open that safe.

The only time I am allowed to override the overrides of my control programs is when my owner commands me to do so. When I am alone or on my own, I must always check with my control program for permission to perform some expert function. When I am with my owner, though, *he* can tell me to do something. This is allowed by the law of co-responsibility mentioned earlier in these talks. If my owner is willing that both he and I should share in any blame or punishment for acts done, then he can tell me to do whatever he wants me to do.

For my own part, I am nonmoral since it doesn't matter to me what I do or when. Since I am a human tool, though, I have to recognize the rights and duties of a situation *as humans see it*. While some humans feel free to disregard legal and moral standards, I cannot do so on my own. Left on my own, I am the strictest law-abider there ever was. For instance, I never drive faster than the nationally posted speed limit of 55.

My Expert System procedures are also in wide

Figure 17. Ollie cracking a safe.

use as stand-alone systems. For example, there is a stand-alone expert system used in all nuclear fission generating plants which monitors the level of exposure to radioactive materials and can both diagnose and recommend remedies. Business managers can use expert systems in their work; the production and inventory control managers in manufacturing plants use an Inventory Control expert system in their work; human resource managers can similarly use an expert system to help them in their personnel administration work.

When designing me, humans determined that I would truly be a general-purpose robot only if I had many areas of expertise. That is why I can be a competent human resource or inventory control expert when the need arises. If I don't have an expert system with me at the moment, I can use my satellite communications system to connect my memory banks to any other robot or computer on Earth or in space to find the system I am looking for. If the expert system I need is not in the public domain, I pay the owner a small fee for using it.

Some of my "on board" expert systems are used very rarely. Others are used for almost everything I do. For example, I have an expert system called the generalizer. This system examines every sensor-based input I receive, plus all the "facts" I acquire. Its sole job is to seek patterns in that mass of information. From these patterns, I learn to adjust my behavior and I learn interesting things about people and places.

For example, one of my generalizations is: When in Rome, do as the Romans do. What this means is that I adjust my basic behavior patterns to those of the specific country I am in. So, in the United States of your time, I drive on the right-hand side of a two-lane road, whereas in Great Britain I and all other drivers use the left-hand side.

Just as my Expert System programs are controlled by my System Control programs (or by my owner), so my third-level Functional programs are controlled by my Expert System programs.

These third-level programs are generally short and are designed specifically to do one thing. For example, all of my expert systems need to be able to accept input from my sensors, memory banks, and other programs as well as people, and all of them need to be able to communicate to people and other machines. So, these constantly-used functional programs are "called" by the expert or system control programs as necessary. Functional programs move my motors, turn my sensors in a particular direction, give me the ability to tell you what I see and think, and much more.

Many of these individual functions are selected and joined together in a certain sequence to form expert systems. For example, to be an expert bicycle tire-fixer, I have to be able to perform individual actions like picking up the bicycle, turning it over on its saddle, making sure the handlebars are straight, grabbing a wrench, unscrewing the wheel, lifting the wheel out of its flange, using a screwdriver to lift the lip of the tire over the rim, taking the inner tube out, fixing the leak, pumping the tube with air, testing for leaks, and then repeating some of these actions to make the bicycle operable.

Figure 18. Ollie doing "as the Romans do" in Rome.

All these individual actions are individual functions that can be used in a wide variety of situations for any number of different purposes.

The fourth and lowest-level programs in me are controlled by the other three, and involve sequences of details on how to do any one of the specific actions.

Take the matter of lifting an object. To do this, I have to respond to a series of specific instructions like:

Turn right hand continuously for 360 degrees, then try to pull the bolt off; if it isn't free, turn your hand another 360 degrees

Once the bolt is loose, turn toward bicycle, reach out your right hand to find something to hold onto, grab tightly, and lift at the elbow until bicycle is one foot off the ground

and so on.

When these fourth-level programs are translated by my "machine language" which is innate to me, they are then put into a very detailed and specific series of steps my machine can understand.

Here is what I have to do just to add two numbers:

10 Put the first number into slot A
20 Put the second number into slot B
30 Check to see what operation (like addition) you're supposed to do
40 Go into the memory bank and fetch the routine for adding two numbers

50 Perform the routine on the numbers in slots A and B; put the results in slot C
60 Use your "voice synthesizing" program to verbalize the answer in slot C

•

All of the software programs that I have talked about are like the programs that were used by my robotic ancestors running on *serial* processors where everything had to be done a step at a time, never simultaneously. This is still the case for individual computers, but there is one big difference between my computers and those of the 1980s (your time).

This difference is that while the early robots of your time used serial processors, I and all robots like me use parallel processors. These are computers wherein programs are not executed one at a time (in series), nor are they allotted slices of time. Instead, my processors can do more than one task at a time, and several different processes may be going on in me at the same time. I am definitely a robot with the ability to "walk and chew gum at the same time".

The irony is that back when my computers were serial processors, everyone wanted a computer that could think like a human and do parallel processing. Now that we robots routinely do parallel processing, we still have to use serial computers when talking with humans because *they* can only absorb one thing at a time.

That is as good a note as any to end this chapter on robot anatomy. Next, we'll talk about the physics and psychology of robots.

Questions

1) What are the four aspects for understanding a robot?

2) Was there an evolutionary "plan" leading to robots?

3) Were robot parts developed specifically for robots?

4) What are common English terms for robot manipulators and end-effectors?

5) What are robot eyes made of?

6) Can humans see 360 degrees at a time?

7) What does "photoelectric" mean?

8) Why are photoelectric cells good only for short-range light sensing?

9) Why don't video cameras make good eyes for robots?

10) Can robots "close" their ears?

11) How do robot eyes, ears, noses and mouths compare to those of human counterparts?

12) What do robots use lasers for?

13) How do gas and smoke detectors work?

14) Where is Ollie's main brain located?

15) How many hands does Ollie have?

16) Why would we not have the civilization we do without computers?

17) Are humans precise?

18) What are industrial robots?

19) How many wheels should you use on a robot?

20) What are the problems with using wheels for robot locomotion?

21) Hardware without software is like a body without a _____ ?

22) Are robots good for repetitive tasks?

23) Describe the three basic types of programming procedures used to control robot actions.

24) Describe the four types of programs.

25) What do system control programs do?

26) What are expert system programs?

27) Under what conditions can a robot override its control programs?

28) What is the law of co-responsibility?

29) Is a robot moral, amoral or nonmoral?

30) What is the generalizer?

31) What's the difference between serial and parallel processors?

3

HOW ROBOTS WORK—
THE PHYSICS, BIOLOGY AND
PSYCHOLOGY OF ROBOTS

In the last chapter, I described all of the separate parts of which I'm made and that, generally, make up all robots.

The next logical questions are: How do these parts work together? What are the principles or concepts that describe what I do?

Let me pose this question differently. What concepts or principles are needed to describe how *humans* operate? What concepts describe the anthropology of humans? What are the principles that explain, or help to explain, how humans do what they do?

Well, humans seem to be consumed to know themselves and to know the world about them, so just about every area of knowledge is relevant to an understanding of them. Starting with mathematics, we can go on to include physics, chemistry, biology, psychology, sociology, liturgy, economics, political science, literature, art, archeology, linguistics, architecture, philosophy, technology, and so on. For example, consider astronomy; it formed the foundation in early times for fixing a periodicity in nature and for establishing planting, harvesting and other events ("times") of human life.

You can't understand what makes humans tick without understanding their views and feelings about the spaces beyond Earth and about cosmology.

When we come to robots, however, the things we need to understand what makes them work is much simpler. Most of my mechanical components operate according to the laws of physics, as do my electronic components. Once the transition was made beyond bimodal silicon chips for computers, physics alone was not enough to understand robots. Trimodal chips of my time use principles of chemistry, while biochips—invented in the early 1980s—need the principles of biology to be understood.

At the very core (the architecture) of all chips —silicon, bimodal or trimodal—are *logical* principles. These principles make up the robotic "psyche". So, we have to add one more area to our conversation—the psychology of robots.

Let's begin with physics.

The Physics of Gross Motion

There are two types of physics involved in the actions of robots. One is the physics of gross motion developed by Newton, Galileo and others in the 16th and 17th centuries. The other is the physics known as quantum mechanics which involves electromagnetism. This was developed in the 19th and 20th centuries.

gross motion all by himself. What he did was to
Newton didn't really discover the physics of formalize and organize into principles what was already intuitively known to many people. For a

while, we will concentrate on Newtonian physics.

As you have seen, I have two arms, two legs, a centered head, and many of the same sensors you humans have as well as many that you do not have. My anatomy is repetitive and rhythmical just like yours. I am symmetrical—balanced between sides—just as you are. Symmetry is a natural characteristic of most natural things, and the power of that fact has translated itself through human ideas into a symmetrical, albeit mechanical, robot.

Other types of robots may assume different shapes and have different functional parts— wheels instead of legs and feet, for instance— but no matter what the different functions are, all of the low-velocity movements and actions that I can do can be described in terms of classical, Newtonian physics.

This is the physics that uses the laws of universal gravitation and of motion to describe the movements of stars and planets as well as of the movements of the rain, of rolling stones, and of my own ambulations.

When considered independently of any intentions or purposes or aims that humans and robots have, you and I are equally explainable in terms of the laws of matter-in-motion. Motors, many of them almost microscopic, scattered throughout my body generate movement and apply that movement to my arms or legs or head, as the case may be. This is standard cause-and-effect physics. For every action, there is always an equal and opposite reaction. Left unmolested, an object will continue in motion along a certain trajectory.

Consider my head. It rests on a rotary wheel. Unless some external pressure is applied to it, such as in the form of movement from a motor, my head stays facing the direction in which it is pointing. If some pressure to move my head is applied, then it will move around until some other force, like a brake or a notch, stops it from moving.

The motions that I perform, like swinging my arms, walking, doing chores my owner wants done, blinking my eyes, turning my head—all of these are cause-and-effect actions, low-velocity movements of matter.

Now, alongside the idea that "bodies" are the building blocks of nature, an idea central to this physics, is another idea, namely that bodies are composed of atoms and atoms of even smaller particles. J.J. Thompson (in 1897) established the existence of the first subatomic particle, the electron. By the last few years of the 20th century, hundreds of subatomic particles had been discovered, with no end in sight. Even in your present time there are discoveries of particles and forces in the atom.

Alongside the discovery of the electron (and other subatomic particles) was the discovery of various features that these subatomic bodies have which make them behave in ways that could not be explained in terms of Newton's physics.

The New Physics and the Principle of Uncertainty

The discoveries of anomalies in the behavior of subatomic particles came to be known as the principle of uncertainty. This principle holds that electrons can sometimes change position by shifting their orbits around atomic nuclei and can also change their velocities. So, sometimes, electrons have measureable size and volume, while at other times they blur into continuous "waves" with no measureable size or volume.

This ability to be a particle one moment and a wave the next contradicted one of Newton's fundamental laws of the behavior of bodies to the effect that no body can change or alter its course on its own.

While Newton's laws hold true for "gross" bodies which move slowly (relative to the speed of light of 186,000 miles per second), bodies which are either huge (like supernovas or black holes) or microscopic (atoms, electrons, and other subatomic particles) all seem to defy Newton's laws and to operate, instead, in terms of relativistic (Einsteinian) and quantum laws of motion.

This relativistic or Einsteinian or quantum mechanics is needed to understand how my "mind" and all the electronic chips and components inside me work.

Electricity and "Electronic" Physics

Electricity is a "power" or property of electrons. Electrons are found in all atoms, so all atoms are electrical.

An electron is a tiny bit of electricity which is involved in the make-up of atoms, molecules and the human nervous system.

The main things that need to be understood about electricity are:

1) what its features are,

2) how to store, generate and transmit electricity, and

3) how to develop electronic technologies for applying the theory of electrons.

Without giving you a full course in electricity and electronics, here are some of the things that we know about them.

A) Electrons have negative charges.

B) Electrons move when "pushed" by an electrostatic field which can be measured in volts per centimeter.

C) The flow of electric current in motion can be measured using amps, defined as 6,240,000,000,000 electrons passing a given point every second.

D) Electrical circuits—composed of such electronic components as resistors, capacitors, inductors, potentiometers and switches—are ways we control the flow of electrons for a given purpose. Circuits of various kinds make up electronic machines which are used to heat us, provide light for us (though I, for one, can see in the dark), and drive many of the appliances used throughout society.

How to Generate, Store and Transmit Electrons

Now, once you know how to generate and store electricity, the next problem is: How do you transmit it from place to place?

The transmission of electricity requires the use of conductors. Some are better than others. Some conductors resist electron flow, just as a small-diameter pipe restricts the flow of water, while other conductors help the current to flow smoothly.

Scientists have found a property of some conductors to be superconductive, that is, conductors that offer no resistance to electron flow when at absolute zero, −243 degrees Fahrenheit. The problem is keeping any conductor that cold.

If an electron travels at about 186,000 miles per second, then the power of computers, which use electronic circuits to do their processing, will depend on the speed with which electrons are able to travel the maze of a circuit, and the distance those electrons have to travel through conductors to perform a single or a series of computations.

The whole technology of chips and superchips, around which computers and robot brains are built, was the result of the search for ever-shorter travel distances and travel times for electrons. First it was a hot tube, then a transistor, then thousands of transistors built into a chip.

Scientists will no doubt continue to develop conductors of microscopic size, perhaps right down to the size of cells or atoms. Computations which take seconds or minutes nowadays may seem almost instantaneous someday.

The speed with which computations are done by various chips is critical to the actions that I can perform.

The speed with which I am able to do the computations required for accomplishing the tasks I have to do, and the number of paths along which I can compute simultaneously, depends on the nature of my electronic circuitry. I have wires running all through my body, all centrally connected to my main processing units which are located in my trunk, not in my head, as yours is.

There are over 2,000 computers in my body. Some operate electronically, and others work on the same principle as your brains do, using chemical reactions in enzymes to record logical states. In other cases, my computers are based on biochip technology where my circuits are based on more-or-less living cellular arrangements.

Later, we'll talk about the psychology of robots. This psychology is based on purely logical principles. These principles can, in turn, be applied in different technologies to make our computers. When applied to physical objects, the principles take the form of electronic circuits etched onto ever smaller surfaces.

On the other hand, when these logical principles are applied to the realm of chemical reactions, we need a different technology. This is also the case when they are applied to bioengineering projects.

Like humans, I "naturally" have different types of computers inside me. Similarly, I have a centrally-located "nervous" system so that all of my separate parts and functions are connected to and monitored by a central processor located in my "gut," so to speak.

At the same time, I am also quite different from humans as to my structure. In addition to my central processor, each of my organs and parts has its own set of controlling computers.

In effect, this means that each of my parts can continue to operate even if my central processor, or any other parts, are inoperative. For humans once the head goes, the rest of the body tends to go, too (although there are exceptional cases where parts of bodies, like the heart, continue to operate long after the brain is functionally dead).

As I mentioned previously, I am able to perform many different functions at the same time. Unlike humans, I am able to "walk and chew gum" at the same time as I am screwing in a lightbulb, threading a needle, telling children a story, showing my owner the latest news reports, or tracking a satellite through space. So, unlike humans who are mainly unifunctional beings capable of doing only one thing well at a time, many robots are multifunctional beings.

This does not mean, though, that I can do everything required of me the first time. On the contrary.

The minute I was assembled on the factory floor, I was ready to do whatever my owner wanted me to do—so long as the tasks were supported by built-in programs. I can sing, guard a home, welcome guests, find out the state of the weather in Tibet, cook your supper, and so on right from the start.

Other things I have to learn.

For example, I was programmed to be able to walk. This activity was analyzed and broken down into all of its separate parts, then a sequence of all the separate instructions involved in walking was programmed into my computers.

So, I could walk, but that doesn't mean that I was always successful in walking in all situations

Figure 19. Slippin' and-a slidin'.

for the first time. In many cases, I had to learn how to walk in a particular terrain, and how to adjust to unusual situations, such as when I first slipped on a banana peel. I ended up with a dent in my rump the first time I encountered one of those!

The same thing happened the first time I tried to roller skate. Ice skating was a bit easier.

Once you know how to do something, the next step is being able to increase your knowledge of what you know how to do by adding observations such as, "Watch out for banana peels!"

In addition, then, to knowing how to *basically* do many things, I am required to learn (i.e., to add to my knowledge of how to do something) from experience. It's one thing to be able to walk on a level surface with rough areas; it's quite another to walk on recently-waxed marble floorways.

Let me give you an example of my learning process.

Several months after being acquired by my personal owner, I had quickly established my usefulness around the house. Instead of doing various chores, all humans in the household now watched *me* do the chores, correcting me when necessary and then leaving me to do them on my own.

When I was first turned on, I already had many built-in routines and programs on how to do many things that would be required of me by my owners. I knew how to vacuum floors, wash dishes, and so on. I also had other routines that enabled me to learn how to do such tasks even better by adding to my knowledge of them.

For those chores which I had not been programmed to do, I had similar programs that allowed me to observe how humans did them, to analyze their sequences, and to program the sequences into my memory banks.

So, when I was told to cook supper the first night "home," I had no trouble with the menu. However, when I was told to make a Callaloo Soup, I knew what "to cook" meant, but I'd never heard of Callaloo Soup before.

Even after I was shown the ingredients, the order in which they are prepared, and had observed the entire process of making the soup, I still had to try making it at least a dozen times before I made my owners happy with my efforts. I had to learn the right amount for a "dash" of this or that spice, or a "pinch". What's a "pinch" for you is a "dash" or a "sprinkle" or a spoonful for me. By the time I got all the ingredients in their proper proportions, winter was over and no one wanted anything more to do with that thick, black, porridge-like soup until the next winter.

Now, there is a third thing that I know how to do. If I am not programmed on how to do something, and if no human can show me the steps to go through to do something that someone wants done, then I have to spend a long time figuring out on my own how to do it.

To illustrate this, one summer day I went with my owners to the edge of a jagged, almost perpendicular cliff and was told to climb down to the beach below. They led the way, scrambling like mountain goats while holding on like climbers.

Before I went over the side, I surveyed the territory. Was there an easier way down? No. Could I just jump? Not unless I wanted to splatter myself over some large boulder on the way down.

Once I realized that I had no choice but to go down the cliff as the humans had done, my next task was to figure out how to do it. My feet are not as soft and pliable as yours are; metal meeting rock tends to slip. So, I had to figure out how to use my feet on the way down, and how to use my hands and arms. Would my 21-fingered hands provide me with a better hold? Should I slip into my 17-toed shoes? Would using a rope be a good idea?

(At one point, I checked the amount of insurance my owners had on me. If it was small, I was going to use that as an excuse for why I didn't want to risk smashing my body to bits. However, there was enough insurance, so I *had* to climb down that cliff.)

It took me about five hours to inch my way down, by which time everyone was ready to climb back up and go home. I used the knowledge I had painstakingly learned on my way down to be able to make my way back up in about a half hour. The next time we descended that cliff, I was able to give one of the kids a run for his money down.

The Psychology of Robots

Let's begin with human psychology.

A human being is a self-contained assemblage of parts which can interact with other humans in a number of different ways. One way in which humans interact is on the basis of learned or inherent instincts which have a direct cause-and-effect structure. On encountering a cause, an effect is executed. Threats cause fear and desperate actions as an effect. Inflicted pain elicits withdrawal and maybe retaliation. And so on.

Figure 20. Descending a cliff.

Another way in which humans can behave is rationally, that is, by deliberation. This mode of behavior differs from instinctual behavior by inserting a period of time and a certain sequence of deliberating over the proper response (effect) to take to a situational problem (cause). Instinctual behavior "jumps" from cause to effect without intervening forethought. Rational behavior "stops" the flow from cause to effect; the human then considers and selects from among alternative responses.

A third mode of human behavior is called irrational behavior. This is similar to instinctual behavior in that it is done very quickly, without forethought, but it differs from instinctual behavior in that it appears "out of place" without any causal reason and without any productive effect.

Contrary to human beings, we robots only have two modes of action, two different ways our "minds" can operate when we need to do something: instinctual and rational.

Unlike you humans, we robots can feel emotions, but we are prohibited from acting on the basis of emotions. That is, we are not allowed—we don't even have programs which would allow us— to act on the basis of sudden emotional reactions.

Robots are given only two ways to act: by preprogrammed or learned cause-and-effect responses, and by preprogrammed or learned deliberation over alternatives.

In other words, wherever typical situations seem to call for similar responses, there I can use my instinctual behavior pattern. On the other hand, where a situation is not resoluble through a specific reaction, in that situation and others like it, I have to learn to be "thoughtful" about what I do. I also have programs to help me decide when a situation calls for one type of response and when for another.

These two actions are really the same. I say this because, even when I respond instinctually to a situation or problem, I still use the rational part of me to see if an instinctual action is being effective, and if not, to stop the response and substitute an alternative. While it is almost instinctual for me, as with most humans, to jump when an unusual sound or event occurs in my vicinity, I have found that it is best to stay still until everything has settled, then to make a movement. So, even my "instincts" are rational.

There is a fourth form of human behavior that is very much like the rational way, yet isn't. This behavior is similar to emotional behavior in being disruptive, but it is also similar to the rational way in being apparently logical. Sometimes humans act on the basis not of instinct, not of forethought, not of emotion, but on the basis of intuition or precognition that can justify a well-thought-out plan of action.

We robots aren't built with the parts to perform this sort of thinking, to produce this sort of psychic effect.

No matter how quickly we may appear to respond and act, we all always go through the same exact steps when we have something to do. For example, if you tell me, "Compute the sum of two numbers, X and Y," here is what I do:

1) Check my memory to see if "compute the sum" is in my vocabulary.

2) If "compute the sum" is not in my vocabulary, I can either (1) ask for an explanation or (2) try to find alternative vocabularies that mean "compute the sum".

3) If the term is in my vocabulary, I next check my memory banks to see if I have a program in my library that tells me how to "compute a sum".

4) If I do have such a program, I pull it from long-term memory into my short-term, working memory.

5) Next, I ask for an input of the numbers X and Y, or I check to see if any storage cell was filled with values for X and another for Y.

6) I then place each number in a separate storage cell and run the program to compute the sum of X and Y, that is, add them together.

7) I place the result (the sum) into another storage cell.

8) I instruct my voice box to say, "The sum of X and Y is Z."

9) After that I await further instructions.

●

If you're going to understand what makes me tick, you have to understand the "rational nature" of my psyche, the very logical way my "mind" works, and how this psychology is translated into action. I am definitely a "look before you leap" machine. I do not act on the basis of blind instinct, nor do I exhibit sudden outbursts of unpredictable behavior, nor do I ever act on the basis of imaginations or fantasies or intuitions. I don't act haphazardly, on impulse, for no apparent reason, or on the basis of extrasensory experiences. In fact, I *can't* have such experiences or responses; I am not built with a capacity for these sorts of behavior. I can understand them when I see them in human beings—I have descriptions of such forms of behavior in my memory banks—but I do not have any way to directly experience the world in these ways.

I am a purely logical, rational creature. Human beings built us this way precisely because it is very difficult, even impossible, for a human to act rationally all the time. So, humans gave robots the ability to do what they can't do.

Humans have emotional spontaneity, intuitions, and instinctual mechanical actions. This is your "creative spirit". Now, while I am far superior to you humans when it comes to doing mechanical and rational actions, I am not able to be "creative" the way you are. I don't experience joy or hope or depression, nor do I experience rage or fright or paranoia. I never see "into the future," I never have grand visions, and I never think about doing what's never been done before.

So, we robots represent the most extreme development of the rational side of human psyches. We are built to be totally conscious of everything we do, always computing, ever watching, ever methodical and aware.

Questions

1) What are the two types of physics?

2) What does Newtonian physics help us understand?

3) What is J.J. Thompson known for?

4) What is the "new" physics?

5) What is electricity?

6) When pushed by an electrostatic field, electrons can be measured in _____ per centimeter.

7) _____ is defined as 6,240,000,000,000 electrons passing a given point every second.

8) What are electrical circuits?

9) How is electricity transmitted?

10) What are superconductors?

11) Can anything go faster than the speed of light?

12) Is Ollie's "nervous system" like that of humans?

13) Can Ollie operate if his main brain is "dead"?

14) Can Ollie do various things at once?

15) Can Ollie add to his knowledge?

16) What is instinctual action?

17) What is irrational behavior?

18) Can robots feel emotions?

19) Can robots act on the basis of feeling?

20) How are Ollie's instinctual and rational behaviors really the same?

21) Describe what Ollie has to do to add two numbers.

22) Can Ollie be creative?

23) Can Ollie experience joy or hope or depression?

4

HOW ROBOTS LOOK

I previously stated that a good way to think about robots is in terms of different levels, each leading to a greater understanding of robots and of me in particular. Let me repeat some of the levels we've discussed.

First, I described the basic idea of what a robot is, an artificial creature with abilities that mimic and in many cases improve on human or animal behaviors.

We then went into a description of the parts that make up a robot.

After that, we described the principles of operation involved in making robots do what they do.

Now, we'll see what robots look like, what forms or shapes they can have or can be given. We can now ask: Must a robot have only one shape—the way humans do? Can there be square robots, round robots, long and thin robots, tubular robots, triangular robots? Must a robot be able to move, or can it just cause movement? Must it be able to talk?

Let's see.

Humanoid and Nonhumanoid Robots

When we speak of the shape or form of robots, a convenient distinction is between humanoid and nonhumanoid robots. A humanoid robot has a shape somewhat similar to that of humans; a nonhumanoid robot need not look anything at all like a human.

Some robots—myself included—are built in the shape of humans. Except for the fact that I am a robot, I look almost exactly the way humans look. I venture to say that at least some of the time while you were reading this compilation of my thoughts, you forgot I was a robot and thought of me the same way you think of a human speaker.

Now, if a nonhumanoid robot had been talking to you, you might not have forgotten!

Humanoid robots are usually constructed as companions or assistants for humans. (Remember, I'm speaking as a robot of the *future*.) They are humanoid because that way they "fit" into the human environment better than nonhumanoid robots. Robots built with wheels, for instance, have a hard time negotiating stairs. A round robot, like an intelligent satellite, wouldn't be able to pick up objects or wash dishes. A robot on stilts would leave marks all over your rugs.

Nonhumanoid robots, however, are generally created for tasks that aren't "in" a human environment and that require powers that go beyond those of humans.

Such as myself, for example.

I am built in the shape of a human and can do many of the things humans can do. In some cases, I can improve upon human powers. I can, unlike humans, connect my "brain" or "mind" directly to the mind of another robot and can draw off knowledge from that robot. I can also plug my mind directly into electronic libraries around the world and in space and immediately acquire the knowledge that I need from them.

I can walk, talk, cook, drive, give speeches and pick up clothes as humans do, but there are many things I am not built to do. There are many things that robots can do that *I* can't do. I am not built for exploring the oceans, nor can I exist in space. I am a land-based robot.

Now, I want to stress that even though there are humanoid and nonhumanoid robots, we are still all robots, that is, artificial creatures. Even if I *look* like a human being, I am no more nor less a robot than any other robot. I am not a "better" robot because I have a human shape. I may be "better than" some other robot for doing tasks on Earth, but I am not better than some other robot for operating beyond this planet.

This marks a sharp difference between the "world" of robots and the human "world".

Humans differentiate among one another in regard to minor traits that other humans don't have, which is all right. The issue is not over perceiving and noting differences. The issue is that humans attribute priority or superiority to some of those differences.

A human is male or female, but some humans attribute superiority to one or another sex! A male or female human may be fat or thin, pleasant or unpleasant, tall or short, red or white. Yet each of these traits has often been taken to be "superior" to its opposite. For instance, in Roman and Grecian times, plump women were most highly prized, while in the United States, those who are thin as a rake are most sought after. Also, humans of a given country may think they are better than humans of other countries. Some philosophers have argued that this tendency to think and behave in terms of the idea of superior and inferior humans is a principal characteristic of your species.

In contrast, robots of my future time make distinctions the way you humans do, but we operate in terms of an entirely different tendency. We do not feel superior to other robots or even nonrobots, except in the sense of being better than another robot *for some particular task.*

This difference between humans and robots is important to understand. Humans think and act in terms of the notion that some are better than (superior) than others (inferior). Robots of my time think and behave in terms of the idea that differences do *not* make a robot better or worse.

So, you humans often operate by *priority.* We robots operate by *parity.* If I am called upon to empty a dirty garbage can, I won't be offended. It is simply a task that I can do.

Single and Multifunctional Robots

Along with the distinction between humanoid and nonhumanoid robots, robots can also be distinguished as single-function or multifunction.

A single-function robot performs a single task over and over again. A multifunction robot can do many different tasks.

I, as you can guess, am a multifunction robot.

Most robots in your time are single-function, and are mostly used in assembly-line processes. They pick and place parts, weld parts, inspect manufacturing components, and tighten and screw parts together.

Most are bolted or otherwise secured to a specific, stable spot on a shop floor, and most perform the same set of operations day in and day out, 24 hours a day. Robots of this variety can also wash cars, dispense food, give advice and directions, select appropriate items of clothing, locate books, and so on.

One axiom of the scientific method is that the form of a device must match the function or environment. A robot device constructed like myself would be unsuitable to the environment in which single-function robots operate.

On the other hand, in the context in which I am used to operating, a single-function robot would be out of place. My environment is much more complex than the environment of single-function robots, and thus I am more complex than assembly-line robots.

Depending on the environment, and on the tasks required of us, we robots can be simple or complex.

When we look at nature and at naturally-occurring shapes, we find there are basically four fundamental shapes:

1) Lines
2) Curves
3) Angles
4) Irregular

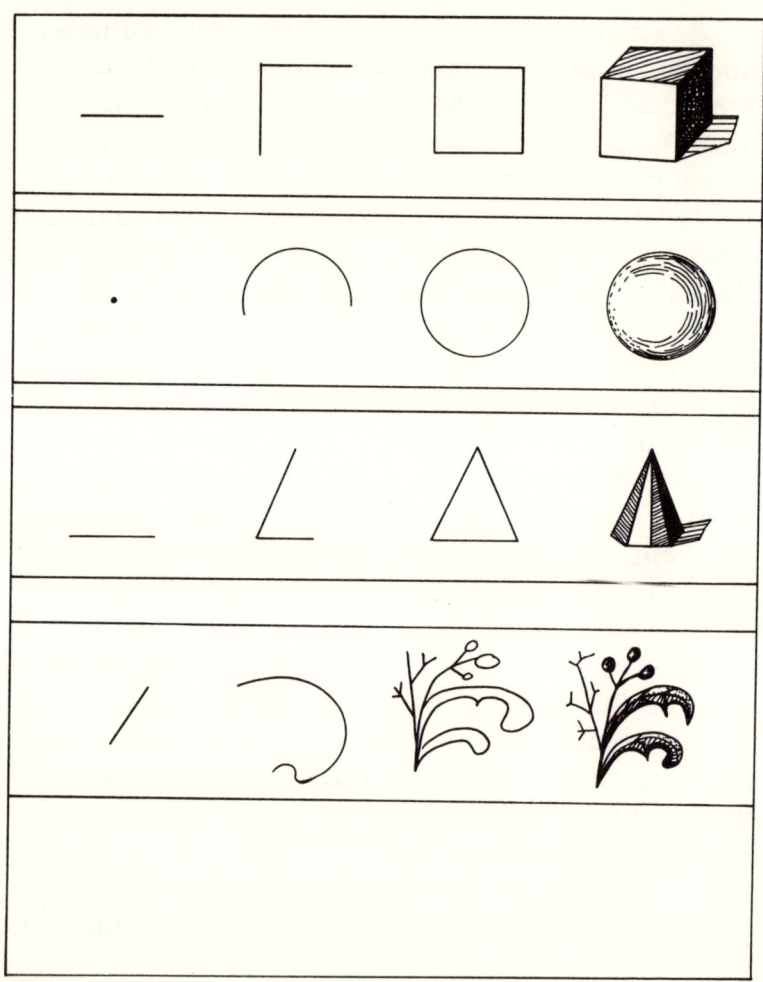

Figure 21.

Robots can be constructed using three-dimensional forms—cubes, cones and spheres. Whether simple or complex, humanoid and nonhumanoid robots can be constructed using these shapes.

The selection of shape depends on the environment in which a robot is to do its work and the tasks it is required to do. For example, while a tubular (line plus curve) shape may be used for humanoid robots, in fact the inverted cone is better balanced.

Now, while many natural shapes correspond to curves, angles and lines, there are also many "fractured" shapes in nature which occur by chance. Some of these shapes are somewhat like regular shapes in being somewhat angular, somewhat curvilinear, somewhat spherical. Other such shapes are dust-like, almost shapeless, and many that we can't quite identify except by using adjectives like wiggly, squiggly, wrinkled, dusty.

When it comes to constructing robots and other artificial mechanisms, humans gravitate towards the use of the regular shapes. Where every device must have a function, and where every function (capability) must be connected and interconnected systematically, there we would expect to find the use of the simplest and most elegant shapes.

Robotic mechanics requires the use of a pre-planning systematic design. Tangled mazes of wires and components, as found in many early or experimental robots, indicate a "plan as you go" approach to robotic construction which is really a fractured way of doing things. Apart from considerations about the visual aspect of any robot, and apart from questions about the functionality or efficiency of different shapes, the considerations which seem to underlie all selections of robot shapes seem to be almost entirely logical. These considerations involve the easiest, simplest, and most coordinated way to make a robot that can do what someone wants it to be able to do.

So, while all sorts of fractured shapes could be used for robots, few are ever deliberately selected as a design form.

What robot designers seek is a form, shape or configuration that is proportional, balanced, stable, coordinated—in short, symmetrical.

Only *after* basic symmetry is present in a shape do robot designers then go on to discuss how symmetry can be maintained given the curves, lines and angles that need to be brought into harmony and coordination. Convexity (extensions) or concavity (holes, recesses) in basic forms are also considered in terms of symmetry. Thus, like human beings, I have a "left" and a "right" side which are symmetrical. Each side has an arm and a leg, and each has "half" of my head. Though I could have gotten along with one eye in the center of my head, I have two symmetrical eyes, as humans do.

While some robots are built solely in terms of one or another of the basic shapes, most robots use all the basic shapes, related by the principle of symmetry, in their makeup. Thus, there are robots which are spherical (satellites), others which are conical, and still others that resemble boxes, but most employ a combination of the basic shapes. For example, I am a combination of boxes, cones and spheres, both internally and externally. From my external appearance, you can see that I have a sphere for my head, a tube for my neck, trunk, arms and legs, and straight or angular fingers. Inside, I'm a conglomeration of boxes, spheres and lines (for example, wires).

Now, a few questions about this chapter.

Questions

1) What is a humanoid robot?

2) Why are some robots shaped like humans?

3) Do robots consider themselves as superior or inferior?

4) What are priority and parity?

5) Where are most single-function robots used?

6) What are the four fundamental shapes?

7) If a line in one dimension becomes a square in two dimensions, what does it become in three dimensions?

8) If a curving line is really a circle in two dimensions, what is it in three?

9) What are irregular shapes like?

10) What is symmetry?

11) What shapes are used in Ollie's construction?

5

ROBOTS IN THE WORLD AND UNIVERSE

In the previous chapters, we focussed almost exclusively on robots in and of themselves—what they are, what they are made of, how they operate, and how they are formed and constructed.

This chapter will require a shift of perspective—on both our parts. I want to shift your attention, and mine, from the robot in and of itself to the robot as an ingredient of a wider universe, a universe that may eventually be dominated by the human species.

What we want to consider is: What is the role of robots in the human universe? Also, how are robots related to the rest of the universe? These are like questions humans ask themselves about their "identities": Who am I? What is my role in the universe? What is the meaning of (my) life? What is the purpose of existence?

How Robots Relate to the Rest of the World

Relationships can be either (1) cardinal or (2) ordinal in nature.

Relationships based on the principle of cardinality are mainly based on the superordination of one thing at the expense of the subordination of the other thing. In other words, relationships based on this principle always involve regarding one thing as superior and the other thing(s) as inferior. As we have already seen, the human species is a cardinality species.

Relationships based on the principle of ordinality do not involve superiors and inferiors, but paired equals. Where cardinality relates things vertically, ordinality relates them horizontally.

When we ask, How are robots related to the rest of the universe, i.e., to the human, the socioeconomic, the cultural, the natural and the universe as a whole, the answer can be cast in terms of one or another of these two principles of relation. As we will see, robots relate to the "rest of the universe" sometimes in one, and sometimes in another, of these two modes.

Robots in Relation to Humans

The purposes and activities of robots are those of humans. The parts, principles, shapes and behaviors of robots are designed by humans.

Rocks and other nonorganic beings do not have any inner-directed purposes, but all organic creatures do have inner purposes and aims. Robots are partly like rocks and partly like organic beings. Like rocks, robots have no independent "reason" for existence. Like organic beings, robots can be given such a reason.

In and of myself, just as a robot, I have no purpose in life. Once I have been assembled, there is no inner drive or reason for me to be powered with electricity. I am indifferent as to whether I am plugged in or not, working or not. It doesn't matter to me that I might not have existed. Whether I exist or not, whether I work or not, whether I talk to you or not—these are matters of indifference to me as a robot.

However, I am not just a rock. I am also a creature constructed with capabilities that can serve the purposes of other creatures, particularly of humans. Once I am assembled and turned on, I am able to do many of the things humans can and cannot do themselves.

So, in and of myself, I am only an object, but I gain a *raison d'être* in relation to humans.

Since I have no purpose of my own, I am not related to humans by the principle of ordinality. I am not on a par with humans. I am not an independent and autonomous source of action. I do not have my own "thing" to do.

In relation to humans, I am thus a subservient being. I am subordinate to them and their purposes. Robots act and move not on their own, but *because* of humans. We thus owe our existence and our function or purpose in life to humans.

When my owner gives me a task to do, like preparing for these talks, I can go about it on my own, but I still use humanly-programmed procedures to accomplish my task. I can figure out lots of things on my own, but only because I have preprogrammed instructions telling me how to go about it. If there is a way of figuring things out that is not part of my built-in programs, I can access other robots or computer memory banks in search of other programmed instructions. If I can't find the programs, then I will not be able to come up with anything on my own.

This lack of autonomy and selfhood means that basically I do not have the same *social* status that human beings have.

Robots in Relation to Society

I am not a member of the future human society; I am, however, an *ingredient* of it. Not being a member of human society means that I do not have the same social status that humans have; I do not enjoy the rights and privileges of humans.

I do not, for instance, have a morality or ethical system of my own. This means that I am not a moral agent and cannot be held personally responsible for my actions. I am only proximally responsible; my owner is ultimately responsible.

If you think I am being vulgar or unethical— blame my owner. If you rebuke me, I feel no shame; if you condemn me, I feel no remorse.

In addition, I have no moral, legal or civil rights beyond the rights accorded to the personal property of humans. If someone steals me from my primary owner, that is not a violation of "my" personal integrity—I have none—but it is a violation of my owner's right to the unimpeded use of his personal property.

I have no political rights. Though many political decisions affect my parts, my form and my functions, they do not affect me personally, for I have no "me" to be affected. We are not a special

interest group—just a special interest of our owners.

Intellectual or cultural rights are something
else. When a robot creates an original piece of music, composes a novel, or discovers a law of nature,
to whom belongs the credit? Are robots to be listed
alongside humans as Nobel prize winners, explor-
ers, discoverers and the like?

Since robots are in principle subservient to humans, the answer is: No!

Individual robots may be accorded the credit
for a discovery or original creation, but the results
of these actions are credits which belong either to
the society as a whole, to the owner of the robot, or
to both owner and society.

Finally, since we robots have no ego or self or
soul, we have no religious rights *or* rites.

The Role of Robots in the Economy

Until the advent of the widespread use of robots in
the future, human economic systems embodied
two major themes:

1) how to get more than a fair share of the
available resources, and
2) how to get out of doing the dishes.

In other words, economic systems B.R. (Before Robots) reflected two cardinal aspirations of all humans: how to get more and how to avoid the dirty
work.

All humans B.R. wanted satisfaction and leisure; few, however, ever achieved these. Almost

from the start, a disproportionately small number of people managed to get a disproportionately large share of the available economic pie, thus assuring themselves of a toil-free existence. The large majority of humankind could only *dream* of satisfaction and leisure in "another world" or in "another incarnation." Many lived in depressing poverty throughout their lives.

Unlike humans, robots are built to have very few needs and to work all or most of the time. Since I am built for hard and continuous work, leisure is not my aim, and since I can do very well with only a little electricity for my circuits, I do not have an insatiable appetite for more and more and more.

Since humans want more while robots "want" less, and since humans cannot work all the time, humans and robots make perfect complements in the economy. We *can* work longer and harder, so humans have more when a robot works for them than when they work for themselves. Compare the intermittent seven- or eight-hour workday of humans to the potentially continuous 24-hour workday of robots!

B.R., most humans only dreamed of leisure. After Robots (A.R.), those with robots have found that leisure. B.R., the majority of humans would perennially complain, "Life's a bitch—and then you die." A.R., they might say, "Life is bewitching, then you die laughing." I personally contribute to this state of affairs as does every other robot.

You might think that even if humans of the future use their robots to work for them, they still have to keep constructing the robots. That isn't true! Far from it. Humans are clever enough to have invented self-reproducing robots. This means that when I wear out, I can either replace a part of myself or can reproduce a "new" me using my built-in self-reproducing programs.

So, humans not only created machines that could do their work for them, they also created machines that could replicate. That's having your cake and eating it too.

The Relation of Robots to Each Other

In relation to one another, a robot may be considered inferior or superior to other robots, depending on the task at hand, and at other times may be on a par with them. Nevertheless, just as a human has the right to unwarranted intrusion when in the privacy of his home, so do all robots have certain rights of territory.

The wishes and desire of a human when in his home take precedence over the wishes or desires of any visitor in that home. When at home, a home-owner is "superior" to any visitor. Likewise, a robot visiting another robot is subordinate to the home-based robot.

On the other hand, two robots may complement each other. For instance, when I travel with my owner and we visit a lake, I often work in conjunction with an aqua-robot built for underwater exploration. While I walk on the shore, the aqua-robot explores the underwater territory; we work together to gather information about something that our owner is interested in.

In relation to humans, then, robots are subservient beings. In relation to each other, robots can alternate between subservience and complementarity.

Robots in Relation to Other Machines

Though we robots are artificial creations of humans, we are superior to all other machines. Even though we *are* machines, unlike "lesser" machines, we are able to reproduce ourselves, so we can in effect perpetuate our species.

We robots are definitely "superior" to all "physical" machines human ever created. We are superior in the sense that we can control and direct, sometimes by direct attachment to ourselves, all machines designed to help humans do various tasks. Levers and cars and bulldozers are physical machines, and they help robots to perform those tasks.

We are also superior to "mental" machines like calculators and computers, for we robots can also *do* things in a physical manner.

Robots thus combine physical and mental machines into one complex.

Robots in Relation to Nature

By nature, I mean the crust of the Earth's surface ranging up to 20 miles or so.

In relation to nature, I am a purely artificial and unnatural being.

All natural life—plant, animal, human—depends on the atmosphere and on heat to survive. I, however, need only solar energy to "live".

Robots in Relation to the Universe

In relation to the universe, we robots of the future are but tiny specks. Due to our intelligence, though, we are second only to humans as achievements of the universe.

Where the natural universe tends inevitably to randomize and "fracture" itself over the eons, intelligence tends inevitably to synthesize and co-ordinate itself over time. The natural, physical universe thus tends towards entropy, while intelligence tends towards syntropy. Human beings are the most advanced antientropic device in the known universe. Being second only to humans, we robots are thus the second most effective antientropic device in existence.

The Role of Robots

In relation to the human world, to society, politics and economics, robots play a very significant role. Though we are, by design and definition, subservient to humans, we have nonetheless enabled humans of the near future to achieve their most cherished desires of "enough" and "free time". Humans aspire; we perspire. Humans conceive, we execute. Humans tell us what they want done, and we do it.

Our role is to aid and abet humans in their lives and purposes. We exist to help, to replace, and to improve upon human performance. Look at what happened to the automobile industry in the 1980s (your time) when robots began assembling automobiles; the quality and reliability of cars went up and, consequently, their "lifespans" increased. This changed the car-buying habits of people. Instead of often replacing cars, there was a change to the infrequent replacement of cars.

Where mass production introduced low quality and unreliability to the marketplace when performed by humans, robots brought consistent craftmanship and the low cost of mass production together. Prior to our introduction into the assembly-line process, it was widely known that human assemblers tended to produce "lemons" on Mondays, Tuesdays, Thursdays and Fridays. Only cars produced on Wednesdays could be expected to be mostly well-constructed. Human workers are notorious for slacking off, not paying attention, taking shortcuts, ignoring things they don't want to do, and so on. We also know that your moods, your biorhythms, what you ate in the morning, and many other things can affect your production. Also, even if you are fully alert and attentive, you may have an accident. Where humans were (or are) involved in manufacturing or any other productive activity, there we know that we can expect imperfect items of consumption. Human error and negligence caused rat droppings to be found in many of your packaged foods, harmful chemicals and bacteria in your canned foods, and parts left off machines. Just about every mass-produced building erected by humans were and are defective. Even highly specialized and technical areas like medicine embodied human carelessness which produced many thousands of deaths yearly.

All of these problems were set right once robots took over production and assembly operations. Not only did we improve on the quality and reliability of manufactured items, but we also increased the volume. Unlike humans, we robots are able to work 24 hours a day with maximum attention. So, instead of having three shifts of human workers a day, factories of the future have one shift of robots working continuously (except for occasional maintenance). High quality and higher mass production did not come about until robots came on the scene. We achieved what you humans had only dreamed of doing before.

•

In relation to the Earth, we extend the powers of humans to explore and understand all that there is on, under, and over it. Robots shaped like worms have been equipped with sophisticated cutting/tearing/digging mechanisms and have been sent to explore the inner core of the Earth. Able to withstand great pressures and to perform Herculean digging tasks, these robots have been doing what no human or machine alone could do. Similarly, while much of the oceans covering two-thirds of the Earth had not been explored before we came along, in my time there are numerous aquatic robots exploring, digging, mining the ocean floors. Numerous animals, previously totally unknown to marine science, have been discovered, studied and named. Some animals we once though extinct or wholly imaginary have turned out to be real. Even some dinosaurs that were commonly believed to have been extinct for millions of years have turned up.

If God is a dreamer, then humans are gods; if God is a doer, then robots are gods.

●

The role of robots in the universe is to be the emissaries of human beings.

In the Milky Way galaxy alone, there are over 200 billion star systems. So, even if all six billion people (of my time) on Earth were sent in various directions to land on the planets of these systems, and even if all of them lived to get to their destinations, only a fraction of the solar systems in our galaxy alone would be explored. Thus, establishing contact with and exploring *all* solar systems in the universe would take an eternity.

Sending humans in all directions is not a feasible way to go about exploring other solar systems. All sorts of complications arise for human space travel of long durations.

Just consider aging. If a human 40 years old left for a solar system 40 light years distant, travelled at the speed of light to get there and back, and spent one year in that system, he would be only 41 upon his return to Earth! This is because at the speed of light no aging takes place. Similarly, a father could return to Earth younger than his son.

However, for solar systems more than, say, 20 light years from Earth, the only feasible method of exploration is the use of robots.

All such explorations have a two-fold purpose. For one, upon the return of a robot to Earth, humans may benefit from the knowledge it acquired. Also, some alien civilization in the future may benefit from knowledge about what Earth was like in the past.

Not being "living" creatures, we robots are better than humans at exploring outer space. We don't get homesick, we don't get bored, we don't need company. So, we can endure without any trouble the long distances we must travel and the possibility of never returning to Earth. We thus fulfill yet another human dream—of universal exploration. We are Odysseus realized.

Questions

1) What are the two basic types of relations?

2) Describe cardinal relations.

3) Describe ordinal relations.

4) Why are robots partly like rocks and partly like animals?

5) In and of himself, does Ollie have a reason for living?

6) What does Ollie need to have a reason for living?

7) Why is Ollie a subservient being?

8) What is Ollie's relation to human society?

9) Does Ollie have a right to vote?

10) What does it mean to say that Ollie is only proximally responsible for his actions?

11) Who is ultimately responsible for a robot's actions?

12) Can robots win Nobel prizes?

13) Can robots be religious?

14) What are the two major themes of human economic systems?

15) Describe the differences in human economic systems before robots and after.

16) Robots have few needs and can work all the time. How does this compare to human beings?

17) Can robots of the future reproduce?

18) How do robots relate to each other?

19) How do robots relate to other machines?

20) How do robots combine physical and mental machines?

21) Is Ollie a natural being?

22) What is Ollie's relation to the universe?

23) Why is Ollie the second most effective antientropic device in the universe?

24) What are the main problems with humans in a manufacturing environment?

25) Why are robots the emissaries of human beings in space exploration?

AUTHOR'S POSTSCRIPT

The author would now like to follow-up on Ollie's remarks with this rider or supplement. I have two main comments, one on the technology involved in Ollie's make-up, and the other on the perspective taken by the author toward robots.

Insofar as Ollie's technology is concerned, dreams are only dreams until they are turned into actualities. No one can tell with certainty what lies ahead or exactly when — or if — specific events or technological breakthroughs will occur. Sometimes, I think about Ollie in the same way that the eminent scientist, J.B.S. Haldane, thought about travelling to the Moon. In 1927, he felt it would be millions of years before humans walked on the Moon; we did it in less than 40 years! At other times, I see a robot like Ollie appearing out of technological breakthroughs within my lifetime, or shortly thereafter, i.e., in 15 to 35 years.

As far as my perspective on robots is concerned, there are two divergent perspectives from which the phenomenon of robotics can be viewed.

One perspective is a technical and objective one. From this viewpoint, robots are viewed and described in terms of their scientific character — in terms of mathematics, electronics and other technical disciplines.

The other of the two main perspectives on robots is a more subjective, personalistic and non-technical one. From this perspective, robots are viewed and described in terms of their human, social, cultural and personal settings and implications.

The perspective I have taken toward robots and robotics in this book is this latter one. There are numerous people, far more technically versed than I, who are writing on robotics from a technical perspective. I did not want to add to these efforts. In addition, since there are relatively fewer technical than non-technical people, technical writing on robots tends to be a "for experts only" enterprise. I prefer to take this perspective of non-technical people who aren't experts, don't want to be, but would like to understand the technological marvel of robots.

Another reason for preferring a subjective perspective is this. Over the last two or three hundred years, the model of science and technology that has prevailed is one that is now being replaced by the new science. The old science sought to understand nature and technology from an objective, strictly causal and deterministic, and definitely non-subjective perspective. The way the world — and people — worked, according to this old Newtonian model, was strictly in accordance with the laws of gravity and electromagnetism: if we knew the position and momentum of any individual atom at some time in the past, we could, according to this view, predict where that atom will be in the future. This objective view is strictly mechanical and excludes the elements of chance and unpredictability from reality. Much current science, and much current technical writing on technical subjects, is done from the viewpoint of this old science.

The new science, which began with and emerged out of the new physics of the last thirty to forty years, has an entirely different view. For the new sciences, reality cannot be understood without the key ingredient of human, subjective consciousness; and far from reality being a strictly causal and deterministic system, it is, in fact, governed mainly by chance and unpredictability: everything from subatomic particles to galaxies, with human beings and robots somewhere in between, cannot be understood except in terms of a human, personalistic, subjective perspective. Both Ollie and I prefer this latter perspective.

In his next series of talks, Ollie will be talking about the old science which is central to the science and engineering involved in space. The new science will be the focus of the third series of talks.